T0312722

Advances in Solid Oxide Fuel Cells VIII

Advances in Solid Oxide Fuel Cells VIII

A Collection of Papers Presented at the 36th International Conference on Advanced Ceramics and Composites January 22–27, 2012 Daytona Beach, Florida

Edited by
Prabhakar Singh
Narottam P. Bansal

Volume Editors
Michael Halbig
Sanjay Mathur

A John Wiley & Sons, Inc., Publication

Published by John Wiley & Sons, Inc., Hoboken, New Jersey.
Published simultaneously in Canada.

For general information on our other products and services or for technical support, please contact our Customer Care Department within the United States at (800) 762-2974, outside the United States at (317) 572-3993 or fax (317) 572-4002.

Wiley also publishes its books in a variety of electronic formats. Some content that appears in print may not be available in electronic formats. For more information about Wiley products, visit our web site at www.wiley.com.

Library of Congress Cataloging-in-Publication Data is available.

ISBN: 978-1-118-20594-5
ISSN: 0196-6219

Printed in the United States of America.

10 9 8 7 6 5 4 3 2 1

Contents

Preface

The 9th International Symposium on Solid Oxide Fuel Cells (SOFC): Materials, Science, and Technology was held during the 36th International Conference and Exposition on Advanced Ceramics and Composites in Daytona Beach, FL, January 22 to 27, 2012. This symposium provided an international forum for scientists, engineers, and technologists to discuss and exchange state-of-the-art ideas, information, and technology on various aspects of solid oxide fuel cells. A total of 57 papers were presented in the form of oral and poster presentations, indicating strong interest in the scientifically and technologically important field of solid oxide fuel cells. The international speakers represented universities, industries, and government research laboratories.

These proceedings contain contributions on various aspects of solid oxide fuel cells that were discussed at the symposium. Fourteen papers describing the current status of solid oxide fuel cells technology are included in this volume.

The editors wish to extend their gratitude and appreciation to all the authors for their contributions and cooperation, to all the participants and session chairs for their time and efforts, and to all the reviewers for their useful comments and suggestions. We hope that this volume will serve as a valuable reference for the engineers, scientists, researchers and others interested in the materials, science and technology of solid oxide fuel cells.

Prabhakar Singh
University of Connecticut

Narottam P. Bansal
NASA Glenn Research Center

Introduction

This issue of the Ceramic Engineering and Science Proceedings (CESP) is one of nine issues that has been published based on content presented during the 36th International Conference on Advanced Ceramics and Composites (ICACC), held January 22–27, 2012 in Daytona Beach, Florida. ICACC is the most prominent international meeting in the area of advanced structural, functional, and nanoscopic ceramics, composites, and other emerging ceramic materials and technologies. This prestigious conference has been organized by The American Ceramic Society's (ACerS) Engineering Ceramics Division (ECD) since 1977.

The 36th ICACC hosted more than 1,000 attendees from 38 countries and had over 780 presentations. The topics ranged from ceramic nanomaterials to structural reliability of ceramic components which demonstrated the linkage between materials science developments at the atomic level and macro level structural applications. Papers addressed material, model, and component development and investigated the interrelations between the processing, properties, and microstructure of ceramic materials.

The conference was organized into the following symposia and focused sessions:

Symposium 1	Mechanical Behavior and Performance of Ceramics and Composites
Symposium 2	Advanced Ceramic Coatings for Structural, Environmental, and Functional Applications
Symposium 3	9th International Symposium on Solid Oxide Fuel Cells (SOFC): Materials, Science, and Technology
Symposium 4	Armor Ceramics
Symposium 5	Next Generation Bioceramics

Symposium 6	International Symposium on Ceramics for Electric Energy Generation, Storage, and Distribution
Symposium 7	6th International Symposium on Nanostructured Materials and Nanocomposites: Development and Applications
Symposium 8	6th International Symposium on Advanced Processing & Manufacturing Technologies (APMT) for Structural & Multifunctional Materials and Systems
Symposium 9	Porous Ceramics: Novel Developments and Applications
Symposium 10	Thermal Management Materials and Technologies
Symposium 11	Nanomaterials for Sensing Applications: From Fundamentals to Device Integration
Symposium 12	Materials for Extreme Environments: Ultrahigh Temperature Ceramics (UHTCs) and Nanolaminated Ternary Carbides and Nitrides (MAX Phases)
Symposium 13	Advanced Ceramics and Composites for Nuclear Applications
Symposium 14	Advanced Materials and Technologies for Rechargeable Batteries
Focused Session 1	Geopolymers, Inorganic Polymers, Hybrid Organic-Inorganic Polymer Materials
Focused Session 2	Computational Design, Modeling, Simulation and Characterization of Ceramics and Composites
Focused Session 3	Next Generation Technologies for Innovative Surface Coatings
Focused Session 4	Advanced (Ceramic) Materials and Processing for Photonics and Energy
Special Session	European Union – USA Engineering Ceramics Summit
Special Session	Global Young Investigators Forum

The proceedings papers from this conference will appear in nine issues of the 2012 Ceramic Engineering & Science Proceedings (CESP); Volume 33, Issues 2-10, 2012 as listed below.

- Mechanical Properties and Performance of Engineering Ceramics and Composites VII, CESP Volume 33, Issue 2 (includes papers from Symposium 1)
- Advanced Ceramic Coatings and Materials for Extreme Environments II, CESP Volume 33, Issue 3 (includes papers from Symposia 2 and 12 and Focused Session 3)
- Advances in Solid Oxide Fuel Cells VIII, CESP Volume 33, Issue 4 (includes papers from Symposium 3)
- Advances in Ceramic Armor VIII, CESP Volume 33, Issue 5 (includes papers from Symposium 4)

- Advances in Bioceramics and Porous Ceramics V, CESP Volume 33, Issue 6 (includes papers from Symposia 5 and 9)
- Nanostructured Materials and Nanotechnology VI, CESP Volume 33, Issue 7 (includes papers from Symposium 7)
- Advanced Processing and Manufacturing Technologies for Structural and Multifunctional Materials VI, CESP Volume 33, Issue 8 (includes papers from Symposium 8)
- Ceramic Materials for Energy Applications II, CESP Volume 33, Issue 9 (includes papers from Symposia 6, 13, and 14)
- Developments in Strategic Materials and Computational Design III, CESP Volume 33, Issue 10 (includes papers from Symposium 10 and from Focused Sessions 1, 2, and 4)

The organization of the Daytona Beach meeting and the publication of these proceedings were possible thanks to the professional staff of ACerS and the tireless dedication of many ECD members. We would especially like to express our sincere thanks to the symposia organizers, session chairs, presenters and conference attendees, for their efforts and enthusiastic participation in the vibrant and cutting-edge conference.

ACerS and the ECD invite you to attend the 37th International Conference on Advanced Ceramics and Composites (http://www.ceramics.org/daytona2013) January 27 to February 1, 2013 in Daytona Beach, Florida.

MICHAEL HALBIG AND SANJAY MATHUR
Volume Editors
July 2012

INVESTIGATION OF NOVEL SOLID OXIDE FUEL CELL CATHODES BASED ON IMPREGNATION OF SrTi$_x$Fe$_{1-x}$O$_{3-\delta}$ INTO CERIA-BASED BACKBONES

M. Brinch-Larsen, M. Søgaard, J. Hjelm and H. L. Frandsen
Fuel Cells and Solid State Chemistry Division, Risø National Laboratory for Sustainable Energy, Technical University of Denmark (Risoe DTU), DK-4000 Roskilde, Denmark

ABSTRACT

Solid oxide fuel cell (SOFC) cathodes were prepared by impregnating the nitrates corresponding to SrTi$_x$Fe$_{1-x}$O$_{3-\delta}$ (STF), x = 0; 0.1; 0.2; 0.3; 0.4 and 0.5, into a porous backbone of Ce$_{0.9}$Gd$_{0.1}$O$_{2-\delta}$ (CGO). STF was chosen as very high oxygen surface exchange rate, high ionic conductivity and electrochemical stability as a thin film electrode have been reported for these materials. XRD measurements showed a high degree of secondary phase formation in the infiltrate as well as reaction with the CGO backbone. Microstructural analysis showed that the STF infiltrate had formed a coating on the CGO backbone. All prepared electrodes were characterized as symmetric cells using impedance spectroscopy. Within the investigated series the infiltrate with x = 0.1 (STF10) showed the best performance with an area specific resistance (ASR) of ASR ≈ 6.4 Ω cm^2 (STF10) at 600 °C in air. The relatively poor performance is believed to originate from poor electronic conduction in the electrodes and possibly also reactions between Sr-containing compounds and CGO. To circumvent the low electronic conductivity, backbones of a composite cathode containing LaCo$_{0.4}$Ni$_{0.6}$O$_{3-\delta}$ (LCN60) and CGO were also tried infiltrated with STF. The ASR of the backbone was 8.7 Ω cm^2 prior to infiltration and this decreased to 0.34 Ω cm^2 after infiltration with STF10.

INTRODUCTION

Rising prizes of fossil fuels and the urge for reducing the global warming call for sustainable and efficient technologies for conversion of chemical energy into electricity and heat. Solid oxide fuel cells (SOFCs) could potentially contribute to a global energy solution. But a number of critical issues need to be resolved before significant commercialization can take place. Issues that must be solved are among others high material costs and high degradation rates. Extensive research of new potential materials is therefore needed for the success of the technology.

In the search for new cathode materials for SOFCs perovskites in the series SrTi$_x$Fe$_{1-x}$O$_{3-\delta}$ (STF) could be interesting since thin film electrodes of the material have shown a performance superior to that of the commonly used La$_{0.6}$Sr$_{0.4}$Co$_{0.2}$Fe$_{0.8}$O$_{3-\delta}$ (LSCF). Oxygen exchange rates in the range of k = 1.2-2.0·10^{-5} cm/s at 800 °C have been reported for compositions x = 0.50-0.95 (STF50-STF95). It is thus expected that the compositions x = 0-0.50 also possess high exchange rates. The purpose of this work is to further evaluate the potential of using STF materials through fabrication and test of porous SOFC cathodes [1].

Previous studies have shown that the bulk electronic (σ_e) and ionic (σ_i) conductivities of STF both increase with increasing iron content throughout the entire compositional range SrTi$_x$Fe$_{1-x}$O$_{3-\delta}$, x = 0-0.99. For SrFeO$_{3-\delta}$ (STF00) σ_e = 20-35 S/cm and σ_i = 0.21 S/cm, whereas for SrTi$_{0.5}$Fe$_{0.5}$O$_{3-\delta}$ (STF50) the conductivities σ_e = 1.0 S/cm and σ_i = 0.05 S/cm have been measured at 850 °C in air [2][3]. The electronic conductivity of STF at 850 °C is thus 1-2 orders of magnitude lower than the electronic conductivity for LSCF of σ_e = 300 S/cm, whereas the ionic conductivity of STF is significantly larger than that of LSCF at 850 °C, σ_i = 0.015 S/cm [4].

1

The thermal expansion coefficient (TEC) of STF increases with increasing content of iron. In the temperature regime ~500-800 °C values of $22 \cdot 10^{-6}$ K^{-1} and $27 \cdot 10^{-6}$ K^{-1} have been reported for $SrTi_{0.05}Fe_{0.95}O_{3-\delta}$ and $Sr_{0.97}Ti_{0.20}Fe_{0.80}O_{3-\delta}$, respectively [5 6]. The ratio of these values is not as expected from the respective compositions. They must thus be attributed some uncertainty. For lower contents of iron values of $17 \cdot 10^{-6}$ K^{-1} and $9 \cdot 10^{-6}$ K^{-1} have been measured for $Sr_{0.97}Ti_{0.4}Fe_{0.6}O_{3-\delta}$ and $SrTiO_3$, respectively [6 7]. The TEC for Fe rich STF compositions is thus much larger than that of usual SOFC components, $\sim 11 \cdot 10^{-6}$ K^{-1} [8], which may cause problems in composite electrode designs.

The need for screening the potential of various STF compositions as cathode materials in a feasible electrode design calls for quick fabrication techniques. It was chosen to fabricate and characterize impregnated electrodes. Hence, an aqueous solution of the nitrates corresponding to STF was introduced into an already physically and chemically stable ionic conducting backbone. The impregnated cell design diminishes the importance of matching the TEC of STF to that of other cell components and allows us to use Fe rich STF compositions without complications. Impregnated electrodes have previously shown excellent performance by impregnation of nitrates corresponding to $La_{0.6}Sr_{0.4}CoO_{3-\delta}$ (LSC40) and $Sm_{0.5}Sr_{0.5}CoO_{3-\delta}$ into backbones of $Ce_{0.9}Gd_{0.1}O_{3-\delta}$ (CGO) [9 10].

In this work the possibility of using STF impregnated porous SOFC cathodes is tested and the most suitable composition is found. STF with compositions $x = 0-0.5$ are characterized with respect to electrochemical performance, phase formation and microstructure. The high contents of Fe have been chosen in order to get high electronic and ionic conductivities, which is desirable in SOFC electrodes.

Two types of backbones have been investigated for the potential of impregnating STF. A backbone of CGO was tried first. This backbone has a very high ionic conductivity and almost no activity towards oxygen reduction. In a second round, backbones of $LaCo_{0.4}Ni_{0.6}O_{3-\delta}/CGO$ (LCN60/CGO) were used. LCN60 has a high electrical conductivity, 1200-1500 S/cm at 25-1000 °C in air, but also a low ionic conductivity as vacancy formation only takes place at high temperature. Although LCN60 itself works as an oxygen reduction catalyst, it was expected that the electrode performance could be improved through impregnation. [11]

EXPERIMENTAL

Preparation of Cells

Symmetric 5 x 5 mm^2 cells with CGO electrode backbones were prepared by screen printing on each side of a 300 μm thick commercial $Ce_{0.9}Gd_{0.1}O_{3-\delta}$ (CGO) electrolyte and subsequently sintering at 1150 °C. The CGO backbones had a porosity of 64 % and a thickness of 30 μm on one side and 45 μm on the other [9]. The differences were due to the screen printing procedure and are not expected to influence the electrode performance, since the electrochemical reaction zones are primarily located within a few tens of micrometers of the electrolyte/cathode interface. The electrode backbones were impregnated with aqueous precursor solutions consisting of a surfactant, strontium nitrate $Sr(NO_3)_2$, iron nitrate $Fe(NO_3)3 \cdot 9H_2O$ and titanium lactate $C_{12}H_{20}O_{12}Ti$ in amounts corresponding to $SrTi_xFe_{1-x}O_{3-\delta}$, $x = 0; 0.1; 0.2; 0.3; 0.4; 0.5$.

The CGO electrode backbones were impregnated according to the procedure illustrated in Figure 1a; (i) – (ii) the porous backbone was impregnated by placing a drop of precursor solution onto the cell and any surplus precursor solution was removed using a tissue; (ii) – (iii) the solvent was evaporated and the surfactant was decomposed by pre-heating in an oven at 350 °C for 15 min. Any leftover infiltrate on the surface of the electrodes was removed, after which a new infiltration followed. This procedure was repeated 14 times.

The amount of infiltrated material was measured by weighing the cells before and after impregnation. Since formation of phase pure STF cannot be expected after calcination at only 350 °C, thermogravimetry of the pre-calcined STF powder was carried out up to a temperature of 1000 °C. Weight losses in the range of 29-36 % resulted from decomposition of the pre-calcined powder.

Subtracting this weight loss and assuming complete formation of STF the resulting volumetric percentage of infiltrated material in the electrode backbones was estimated to 15 % (STF00, density ρ = 5.26 g cm^{-3}) and 17 % (STF50, ρ = 5.15 g cm^{-3}) of the total electrode volume, including pores and backbone. Densities were calculated from molar masses of the constituent elements and unit cell parameters [5].

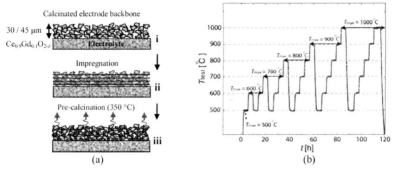

(a)   (b)

Figure 1. (a) The impregnation procedure starting with a sintered electrode backbone on top of an electrolyte. The backbone was impregnated with a precursor solution and pre-calcined at 350 °C. The last step was followed by a new impregnation. (b) Temperature cycle during symmetric cell characterization. The cells were tested at various temperatures (T_{test}) after different calcination temperatures (T_{max}). The arrows show the T_{max} periods of the test.

Symmetric cells with 5 x 5 mm^2 LCN60/CGO 50/50 wt% electrode backbones were impregnated as well. The LCN60/CGO symmetric cells were electrolyte supported by a 200 μm thick CGO electrolyte. The fabrication of these electrodes is described elsewhere [11]. Electrode thicknesses varied between 45-55 μm on one side and 70-85 μm on the other.

The LCN60/CGO backbones were impregnated with the most suitable infiltrate based on the findings from the backbones consisting of CGO, which was found to be STF10. The number of impregnations along with the impregnation procedure was slightly varied for some of the impregnation cycles compared to the impregnation of pure CGO backbones. This was done by increasing the amount of surplus precursor solution left on the cells and by leaving the cells in a vacuum chamber at a pressure of ~ 1 mbar for 5 min. prior to pre-calcination. The vacuum step removes air in the backbone pores and increases the penetration of the precursor solution into the pores [9]. The number of impregnations was varied to optimize the cell performance, and the abovementioned modifications of the impregnation procedure were carried out in order to optimize the amount of infiltrate left in the backbone after each impregnation cycle.

Electrochemical Characterization

The symmetric cells with infiltrate were contacted with Pt paste from Ferro and impedance measurements with in-situ sintering of the infiltrate were carried out using a Hioki impedance analyzer with a frequency range from 0.06 Hz to 100 kHz or a Solartron 1260 impedance analyzer with a frequency range from 0.06 Hz to 1 MHz. The impedance was measured with 15-20 points per frequency decade.

The measurements were carried out on 4 cells at a time at an air flow of 6 NL/h using the temperature cycle shown in Figure 1b. This temperature cycle was applied in order to test the symmetric cells at different temperatures (T_{test}) and after different calcination temperatures (T_{max}).

Initially the cells were sintered at 500 °C for 120 min. and the impedance was measured for one cell at a time. This procedure was repeated at every temperature plateau. The temperature was then increased to the temperature plateau at T_{max} = 600 °C at a rate of 300 °C/h and ramped back down to 500 °C. This procedure was repeated up to temperature plateaus at T_{max} = 700 °C, 800 °C, 900 °C and 1000 °C with a temperature plateau at every 100 °C while ramping up the temperature, whereas the temperature was ramped directly down to 500 °C after each T_{max} plateau. The temperature cycle gives the possibility of comparing the impedance for a cell sintered at a certain T_{max} at various test temperatures $T_{test} \leq T_{max}$, and furthermore the impedance variation with sintering temperature can be followed by comparing measurements recorded at a certain T_{test} but with varying T_{max}. Some tests were carried out up to 900 °C only, since strong degradation of the performance was observed above this temperature. Impedance results were normalized to the cell areas and divided by two.

Characterization of Microstructure and Phase Purity

Impregnated cells sintered at relevant temperatures were fractured by hand and contacted with carbon tape as preparation for scanning electron microscopy (SEM). SEM measurements were carried out using a Zeiss Supra 35 scanning electron microscope.

A precursor solution of STF00 and STF50 was heated to 350 °C for 7 h and subsequently crushed. High-temperature X-ray powder diffraction (HT-XRD) measurements that allows for in-situ calcination were carried out using a Bruker D8 Bragg-Brentano diffractometer with Cu Kα radiation to verify the formation of the STF phase and to study the formation of other phases in the dried precursor solutions. Measurements were recorded for the compositions STF50 and STF00 to retrieve information about the two end members studied in this work. The temperature was varied from 500-900 °C using a Pt heater, heated by Joule heat. Measurements were recorded at every 100 °C. The powder was calcined for 1.5 h (STF50) or 3 h (STF00) at the respective temperature before measuring in order to mimic the sintering times of the symmetric cells before the impedance measurements. Measurements could only be recorded for relatively small amounts of powder at a time, approximately 200 mg. The 2θ range was 17-80 ° with a step size of 0.06 °, and the measurement time per point was 5 s.

The possibility of reactions between STF and CGO was studied using XRD powder diffraction on powder mixtures calcined at 1100 °C for 10 h. Powder mixtures consisted of CGO powder with a surface area of 6.0 m^2/g mixed with dried STF50 and STF00 powder, respectively.

RESULTS AND DISCUSSION

Phase Evolution and Microstructure as Function of Temperature

The results of the HT-XRD are shown in Figure 2a (STF50) and Figure 2b (STF00). The lower XRD pattern in both figures shows a measurement recorded at 30 °C of the powder of the precursor solution heated to 350 °C. Above it is shown patterns recorded at the temperature stated to the right in the figures. The measurement marked as 30 °C* has been measured at 30 °C after cool down from calcination at 900 °C. At higher temperatures the peaks move towards lower Bragg angles due to lattice expansion. Some measurements have been excluded for both powders, since no change in peak intensity could be observed.

From the HT-XRD patterns for the STF50 powder in Figure 2a it is observed that the only clearly detected phases after calcination at 350 °C are those of the Pt-heater (■ identified as PDF no. 00-004-0802), Sr(NO$_3$)$_2$ (✱ PDF no. 00-025-0746) and SrCO$_3$ (▼ PDF no. 01-084-1778). Sr(NO$_3$)$_2$ is not observed at a sintering temperature of 500 °C and above. In contrast, previously observed decomposition temperatures of Sr(NO$_3$)$_2$ are in the range of 550-620 °C [12][13]. This discrepancy may be attributed to uncertainties on the temperature of the Pt heater and the influence of other present phases on the decomposition and formation temperatures. SrCO$_3$ is observed in all the recorded measurements. However, the peaks of the SrCO$_3$ phase appear only weakly at 350 °C and decrease significantly when heating to 900 °C. In a similar experiment performed on a dried precursor solution of La, Sr and Co nitrates the $SrCO_3$ was undetectable at 900 °C and above. This discrepancy is believed to be due to a lower stability of STF compared to the formation of $La_{0.6}Sr_{0.4}CoO_{3-\delta}$ (LSC40) in the temperature range 500-900 °C [9].

In Figure 2a the STF50 phase (◆ identified as $SrTi_{0.5}Fe_{0.5}O_{2.85}$ PDF no. 01-084-1004) appears, although only weakly, at 500 °C, but the peaks increase significantly when heating to 700 °C and 900 °C showing increased but not complete formation of this phase. The exact oxygen non-stoiciometry of the phase is not known, since the XRD patterns of various non-stoichiometries are almost identical. It should be mentioned that the XRD peaks of the isolating $SrTiO_3$ phase are located at almost the exact same Bragg angles as the STF50 phase. The unwanted $SrTiO_3$ may thus be present, although further investigation would have to be carried out to reveal the presence of the phase as well as the exact amount. Cooling down from 900 °C to 30 °C a SrO phase (● PDF no. 00-048-1477) appears.

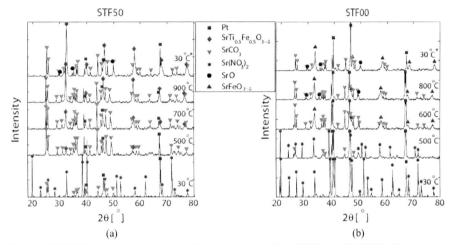

Figure 2. HT-XRD patterns for precursor solution powders of (a) STF50 and (b) STF00. The patterns noted as 30 °C, shown in the bottom of both figures, were recorded after drying at 350 °C, whereas the patterns in the top (30 °C*) were recorded after calcination at 900 °C.

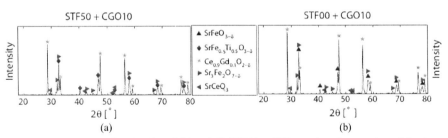

Figure 3. XRD patterns of (a) STF50 + CGO and (b) STF00 + CGO powder mixtures after calcination at 1100 °C for 10 h.

The observed secondary phases do not include iron nor titanium, which must thus be present as additional secondary phases. It has been attempted to match the XRD patterns with the full range of compunds containing either Fe, Ti, Sr, O, C or N but none of these phases were detectable.

The HT-XRD measurements for the STF50 powder thus point at formation of the STF50 phase increasing continuous and strongly between the temperatures 500-900 °C. Furthermore, the XRD data clearly indicate the presence of several unwanted secondary phases. Further investigations are required to reveal the exact amount of secondary phases and STF, respectively. However, comparison with other studies indicates a large presence of $SrCO_3$ in the STF infiltrate and thus a weaker presence of the perovskite phase [9].

The HT-XRD pattern for STF00 is shown in Figure 2b. The $Sr(NO_3)_2$ phase (✳) is again observed after heating to 350 °C and is, as in the case of the STF00 dried powder still present after sintering at 500 °C unlike for the STF50 powder. The $SrCO_3$ phase (▼) is first observed at 500 °C, however the peaks appear only weakly below 600 °C. The phase is still observed after sintering at 900 °C like for the STF00 powder. The STF00 phase (▲ identified as $SrFeO_{2.73}$, orthorhombic, PDF no. 00-040-0906) is first clearly observed at 600 °C. The peak signals from STF00 appear to increase only weakly with increasing temperature. Further phase formation is expected at a temperature above 900 °C. However, as expected the presence of the poorly conducting, ordered brownmillerite phase, $SrFeO_{2.5}$ can be excluded [2 14].

XRD diffraction patterns for mixtures of CGO with STF00 and STF50, which have been calcined at 1100 °C, are shown in Figure 3. The CGO phase (✶ identified as $Ce_{0.9}Gd_{0.1}O_{1.95}$ PDF no. 01-075-0161) is clearly observable in both powder compositions. Furthermore STF50 (♦ identified as $SrTi_{0.5}Fe_{0.5}O_{2.88}$ PDF no. 01-081-0685) and STF00 (▲ identified as $SrFeO_{2.86}$, tetragonal, PDF no. 00-039-0954) phases are observed from Figure 3a and Figure 3b, respectively. The secondary phases observed at 500-900 °C are not observable after calcination at 1100 °C. However, new peaks have appeared; $Sr_3Fe_2O_{7-\delta}$ (▶ identified as $Sr_3Fe_2O_{6.75}$ PDF no. 00-045-0400) and a weak trace of $SrCeO_3$ (◀ PDF no. 00-047-1689), are observed for both powder mixtures. The $Sr_3Fe_2O_{7-\delta}$ phase has previously been observed in various synthesis methods of strontium ferrite after powder calcinations at 600 °C and 1100 °C and has been shown to decompose again after heating for 100 h at 1100 °C [15]. The formed $SrCeO_3$ phase is a stable, electrically insulating perovskite and is highly unwanted as a secondary phase [16]. It is a product of reactions between $SrCO_3$ and CeO_2 and forms at temperatures above 700 °C [17]. This shows that a reaction takes place between the backbone and the infiltrate.

Micrographs showing representative microstructues of a non-impregnated CGO backbone and impregnated CGO backbones are shown in Figure 4. The porous backbone without infiltrate in Figure 4a shows particle sizes in the range 100-300 nm. The infiltrate is clearly observed as small particles of sizes below 20 nm in the STF50 impregnated backbone sintered at 600 °C (Figure 4b). The infiltrate particles appear to be interconnected, which is important for electronic conduction. To some degree the infiltrate also appears to form a thin coating of the CGO particles. and the two components seem to form a good contact. Sintering at 1000 °C implies increased particle sizes of the infiltrate and thus a decrease of the amount of triple phase boundary (TPB), e.g. reduced reaction area, see Figure 4c. After sintering at 1000 °C the STF50 infiltrate is hardly distinguished from the backbone, but it is believed to appear as particles with sizes below 100 nm. The contact angle between the backbone and infiltrate phases appears to be very small, and from the XRD investigations the particles are likely to have reacted with the backbone, forming $SrCeO_3$.

The STF00 infiltrate was significantly harder to detect, although the weight increase due to infiltrated material showed almost equal amounts of infiltrated material of STF00 and STF50. A CGO backbone with STF00 infiltrate sintered at 600 °C is shown in Figure 4d. It is believed that the STF00 impregnated electrodes contain in the vicinity of 15 vol% as calculated from the weight increase due to impregnation. The infiltrate is partly present as clearly observable agglomerates as shown in Figure 4d and a hardly observable thin coating of the backbone particles. The coating appears as small roughness features on the otherwise smooth backbone particles and edges of the coating have been observed. It is preferable that the infiltrate forms a lot of smaller particles instead of the coating in order to maximize the available surface area for oxygen reduction.

The morphological differences between the STF00 and STF50 infiltrates are attributed to a higher sinterability and possibly higher mobility on the CGO backbone of the STF00 particles. The formation of an infiltrate coating has previously been observed for other perovskites on a YSZ backbone [18][19]. The tendency of the STF00 infiltrate to coat the backbone is further evidenced in Figure 4e, where the STF00 infiltrate has been sintered at 1000 °C and does not appear as individual particles unlike the STF50 infiltrate.

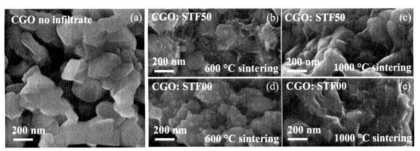

Figure 4. SEM micrographs of CGO backbones with (a) no infiltrate, (b) + (c) STF50 infiltrate calcinated at 600 °C (b) and 1000 °C (c) and (d) + (e) STF00 infiltrate calcinated at 600 °C (d) and 1000 °C (e).

Effect of STF Composition on Electrode Performance

Figure 5a shows Nyquist plots of impedance data from symmetric CGO cells infiltrated with the six investigated compositions $SrTi_xFe_{1-x}O_{3-\delta}$, x = 0, 0.1, 0.2, 0.3, 0.4, 0.5. The cells have been tested in air at T_{test} = 600 °C after in-situ sintering at T_{max} = 600 °C. The insert is a zoom-in of the data for the cells infiltrated with STF00 and STF10. A trend is clearly observed; the data for the STF50 impregnated cell have the highest impedance, and the total area specific resistance (ASR) is decreasing with increasing content of iron. This tendency corresponds to the increase in σ_e and σ_i with increasing amount of Fe in the composition but is also influenced by the differences in microstructure and the amount of perovskite phase present for different STF compositions as elaborated below. The cells infiltrated with STF00 and STF10 appear to perform almost equally well at T_{test} = T_{max} = 600 °C. Including other sintering and test temperatures, the STF10 infiltrated cells generally perform the best at 500 °C < T_{test} < 750 °C, while the STF00 cells perform better at 750 °C < T_{test} < 1000 °C.

The data in Figure 5a show two clearly distinguishable impedance arcs, which both decrease systematically in magnitude with increasing Fe content of the infiltrate. The low frequency (LF) arc is attributed to a convolution of low frequency electrode processes, which are dominated by the electrochemical reduction of oxygen and the migration of oxide ions between the electrode reaction zones and the electrolyte. The LF arcs show very high impedances of R_{LF} ≈ 1.5-15 Ω cm^2 in comparison to similar previous studies for CGO backbones impregnated with LSC40, where a polarization resistance of R_p = 0.04 Ω cm^2 has been reported at 600 °C in air [9]. Hence, the LF electrode processes must be strongly impeded in the STF impregnated electrodes. The impediment of the LF processes are mainly explained by the widespread formation of secondary phases and insufficient phase formation of STF, since it is not expected that any of the observed secondary phases can replace neither the ionic and electronic conductivity nor the catalytic activity of the perovskite.

From the SEM investigations the infiltrate particles are expected to be sufficiently interconnected for electron transport to the TPB (reaction zone). However, a combination of a low electronic conductivity and insufficient phase formation of STF may impede the electronic and to some degree also the ionic conduction in the electrode. $SrCeO_3$ that results from reaction of the infiltrate with the backbone is a very poor conductor with a total conductivity of $1.6 \cdot 10^{-4}$ S/cm at 850 °C in air, and may thus cause a significant impediment of the transport of oxide ions from the reaction zone on the STF infiltrate to the CGO phase [16]. Previous studies did not detect the $SrCeO_3$ phase below 700 °C [17]. However, it is not unlikely that the phase is present in minor amounts. Formation of a few nanometers of the phase at the backbone-infiltrate interface would be fatal for the electrode performance because of the low conductivity of the phase.

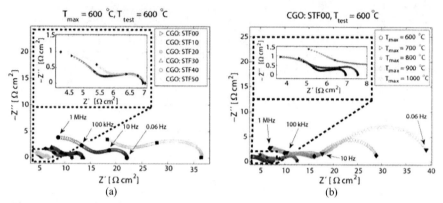

Figure 5. Nyquist plots of impedance data obtained for impregnated symmetric CGO backbone cells in air. (a) Comparison of STF compositions. (b) Comparison of in-situ sintering temperatures (T_{max}) for $SrFeO_{3-\delta}$ (STF00). Characteristic frequencies of 0.06 Hz, 10 Hz, 100 kHz and 1 MHz marked.

The HF arc shown in the plots in Figure 5a are observed at unusual high frequencies (~ 1MHz), at which MIEC cathode electrode processes are not normally observed. The HF arc cannot be due to the electrolyte nor the backbone, as it has been observed even at T_{test} = 800 °C and varies in magnitude with STF composition. The high frequency processes can be caused by either low electronic conductivity in the electrode or formation of the $SrCeO_3$ phase. Widespread presence of non-conductive secondary phases in the infiltrate will lower the net electronic conductivity and may also partly disconnect the electronic percolation in the electrode. Formation of the $SrCeO_3$ at the infiltrate-backbone interface will also lead to a HF-process in the impedance spectra.

It is associated with a significant uncertainty to determine series and polarization resistances of the symmetric cells due to the incomplete HF arcs. The electrode performances are thus evaluated by $ASR_{electrode} = ASR_{total} - R_{el}$, where ASR_{total} is the total area specific resistance of the cells (largest real part of the impedance) and R_{el} is the calculated electrolyte resistance for half the electrolyte thickness. $R_{el} = l_{el} / \sigma_{CGO}$, where l_{el} is half the electrolyte thickness and σ_{CGO} is the electrolyte conductivity calculated from $\sigma_{CGO} \cdot T = 1.09 \cdot 10^5 \cdot \exp(-0.64 \text{ eV}/(k_B T))$ S·K/cm, which implies R_{el}(300 μm, 600 °C) = 1.2 Ω cm². In this way the HF arc is included in the results of R_p. Due to the large values of R_p this procedure only give rise to a very small uncertainty on the determination of R_p. Standard impedance fitting with equivalent circuit elements would not be feasible as the HF arcs are too incomplete. For $T_{test} = T_{max} = 600$ °C in air we thus get $ASR_{electrode}$(CGO: STF00) = 6.3 Ω cm^2 and $ASR_{electrode}$(CGO: STF10) = 6.4 Ω cm^2.

Effect of Sintering Temperature on Electrode Performance

Figure 5b shows Nyquist plots at a test temperature of 600°C after different sintering temperatures between $T_{max} = 600$-1000 °C for a symmetric cell with CGO backbone infiltrated with STF00. The performance of the cell is observed to be strongly dependent on sintering temperature. The performance is best at $T_{max} = 600$ °C and decreases with increasing T_{max}. The optimal sintering temperature is a balance of sufficient phase formation of the catalytically active phase and microstructure. Sintering at higher temperatures increases particle sizes as previously observed but on the other hand, the perovskite phase is not present or only present in small amounts at low sintering temperatures. Furthermore the amount of SrCeO$_3$ may also increase with increasing sintering temperature. From Figure 5b the ASR is seen to increase strongly at $T_{max} \geq 900$ °C, which may be explained by significant infiltrate particle growth and the increased tendency to form SrCeO$_3$ at elevated temperatures.

Impregnation of Electronically Conductive LCN60/CGO Backbone with STF10

Because of the indications of poor electronic conductivity in the impregnated CGO backbones electronically conductive LCN60/CGO backbones were impregnated. STF10 was chosen as infiltrate as it showed the best performance in the impregnated CGO backbones. Impedance results for impregnated and non-impregnated backbones of LCN60/CGO, sintered and tested at 600 °C in air, are presented in Figure 6. The measurements were conducted in the frequency range 0.06 Hz – 100 kHz, since no high-frequency arc was encountered, which is attributed to better electronic conductivity in the electrode. The LCN60/CGO cell without infiltrate exhibits a performance of $ASR_{electrode} = R_p = 8.7$ Ω cm^2 at 600 °C in air. The performance has been evaluated by using the data points measured at 100 kHz and 0.06 Hz.

A large variation in R_s for some of the LCN/CGO cells has been observed. As seen from Figure 6 two of the cells exhibit a very high series resistance, which is attributed to contact resistances at the electrode-electrolyte interface. It is although expected that impregnation decreases R_s significantly for the cells with bad contacting. Values of $ASR_{electrode}$ have although proved highly reproducible.

Figure 6 shows a significant improvement of the electrode performance when impregnating with STF10. $ASR_{electrode}$ of the cell infiltrated 4 times with STF10 has been measured to 0.34 Ω cm^2 and for the cell impregnated 10 times with STF10, $ASR_{electrode} = 2.7$ Ω cm^2, corresponding to electrode performance improvements of 25 and 3 times, respectively, compared to the non-impregnated cell. Hence, the optimal number of impregnations is expected to be closer to 4 than 10 and the number of impregnations has a large impact on the electrode performance. Previous studies by Samson et al. impregnation of a CGO backbone with LSC40 show performance variations of no more than a factor of 2 when impregnating 6, 9 and 12 times and an optimal performance at 9 impregnations [9].

The differences between the present study and that of Samson et al are expected to be caused by the content of the catalytically active LCN60 in the electrode backbone in this work. For illustration of the catalytic activity, LCN60/CGO electrodes with optimized microstructure exhibit a performance of $ASR_{electrode} = 0.4$ Ω cm^2 at 600 °C in air, thus almost as good as the best impregnated LCN60/CGO electrodes in the present study [11]. The LCN60 particles get increasingly blocked with the number of impregnations, which reduces the gas accessibility to the surface and thus the oxygen reduction activity of the LCN60. This hypothesis would mean that the net catalytic activity of the STF infiltrate is less than that of LCN60. This may be due to the presence of secondary phases or less optimistic materials parameters than reported in literature. Further investigations would have to be carried out to confirm this.

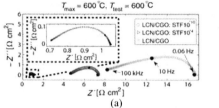 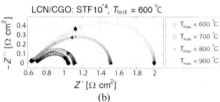

(a) (b)

Figure 6. Nyquist plots showing (a) impedance measurements of LCN60/CGO symmetric cells impregnated with STF10 10 times ($\circ$), 4 times ($\triangleright$) and without infiltrate ($\ast$). The cells were sintered and tested at 600 °C in air. (b) Nyquist plots showing performance variation with in-situ sintering temperature (T_{max}) at 600 °C in air for a STF10 impregnated symmetric cell with LCN60/CGO backbone.

The performance impact of sintering temperature is illustrated in Figure 6b. The electrode performance is best at a sintering temperature of 600 °C. Increasing the sintering temperature decreases the electrode performance, although not nearly with as large a factor as observed for the impregnated CGO backbone. Again, this is explained by the presence of the catalytically active LCN60 phase, in which the particle sizes are only slightly affected by increasing the sintering temperature from 600-900 °C and thus do not imply a performance decrease. However, the significant performance decrease with increased T_{max} shows that the infiltrated STF10 must contribute significantly as a catalyst for the electrochemical reduction of O_2.

Figure 7. SEM photos of LCN60/CGO electrode backbones (a) without infiltrate, (b) with STF10 sintered at 600 °C and (c) with STF10 sintered at 900 °C.

Micrographs of a non-impregnated LCN60/CGO backbone are shown in Figure 7. The non-impregnated backbone in Figure 7a shows large particles with sizes ~200 nm and smaller particles with sizes < 100 nm due to a different particle size distribution for the LCN60 and CGO phases. After 4 impregnations with STF10 and sintering at 600 °C the infiltrate can be seen as particles smaller than those in the backbone, see Figure 7b. Sintering at 900 °C entails agglomeration to particles of sizes ~50 nm as shown in Figure 7c. As previously observed the particles seem to form a very low contact angle on the backbone particles.

The electrode performances of STF10 impregnated CGO and LCN60/CGO backbones and a non-impregnated LCN60/CGO backbone are compared in an Arrhenius type plot in Figure 8. The ASR of the non-impregnated LCN60/CGO backbone varies only slightly with T_{max}, showing limited degradations, since the cells have already been sintered at 1000 °C during the fabrication process [11].

The performance of the STF10 impregnated CGO backbone changes remarkably little with increased T_{max} in the range 500-800 °C compared to the STF10 impregnated LCN60/CGO backbone as well as other cells that have been tested in this study.

The activation energies shown in Figure 8 have been calculated for the data points corresponding to T_{max} = 800 °C. The non-impregnated LCN60/CGO electrode with E_a = 70.9 kJ/mol performs better than the STF10 impregnated CGO backbone at T_{test} > ~700 °C but worse at lower temperatures. The relatively low activation energy of the impregnated CGO backbone is believed to be due to the large influence of electronic and oxygen vacancy conduction limitations on the electrode performance. These processes are less activated with temperature, which can be exemplified by comparing E_a = 61.8 kJ/mol for vacancy conduction in CGO and an activation energy for a thin film electrode of STF50 of E_a = 173.7 kJ/mol. The performance of the thin film electrode is largely dominated by the reduction process of oxygen. [1][20]

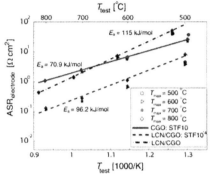

Figure 8. The electrode performance, $ASR_{electrode}$, as function of temperature. The performance has been plotted at various temperatures with a symbol showing the in-situ sintering temperature (T_{max} = 500-800 °C) of the cell. An Arrhenius fit to the data obtained for T_{max} = 800 °C has been included.

The performance of the best LCN60/CGO backbones impregnated with STF10 is still an order of magnitude lower than that of LSC40 impregnated CGO backbones even though the electronic conductivity issues to some degree have been solved by introducing the electronically conducting LCN60/CGO backbone. This may have several explanations; the net catalytic activity and net conductivities of the STF infiltrate are not expected to be very good, the STF particles constitute a significantly reduced surface area compared to the LSC40 infiltrate in the study by Samson et al and the suspected infiltrate-backbone reactions to form $SrCeO_3$ can cause a strong impediment of oxide ion transport from the infiltrate to the CGO backbone. [9]

SUMMARY

$SrTi_xFe_{1-x}O_{3-\delta}$ (STF) was evaluated as potential material in solid oxide fuel cell (SOFC) cathodes by impregnating $Ce_{0.9}Gd_{0.9}O_{2-\delta}$ (CGO) backbones with the compositions x = 0; 0.1; 0.2; 0.3; 0.4 and 0.5. On average the best performance was observed for STF10 impregnated CGO backbones with an electrode area specific resistance (ASR) of 6.4 Ω cm^2 at 600 °C in air. A large part of the ASR is attributed to poor electronic conductivity of the electrode and possibly formation of the highly resistive phase $SrCeO_3$.

$CGO/LaCo_{0.4}Ni_{0.6}O_{3-\delta}$ (LCN60) electrodes 50/50 wt% with high electronic conductivity were impregnated with STF10 (x = 0.1) and showed an ASR of 0.34 Ω cm^2 at 600 °C in air, which is an improvement of 25 times compared to the non-impregnated electrode with an ASR of 8.7 Ω cm^2. It is suspected that the temperatures needed for formation of STF are too high for compatibility with the impregnation technique, which implies extensive formation of secondary phases. Although no precise conclusion for the potential of STF as cathode material in SOFC has been stated, the following has been found:

- If implications of the high thermal expansion coefficient (TEC) can be overcome, STF compositions with a high content of iron such as STF10 show the most interesting properties.
- The formation of STF from strontium and iron nitrate and titanium lactate is accompanied with formation of several unwanted secondary phases, even when heating to 1100 °C.
- STF infiltrate and CGO may react to form an isolating SrCeO$_3$ phase at high temperatures.
- STF infiltrate forms a coating on CGO instead of the wanted high-surface area distribution of infiltrate particles.
- STF infiltrate sintered in-situ at 500-900 °C does not provide the desired combination of high oxygen surface exchange, high electronic conductivity and good ionic conductivity.

ACKNOWLEDGEMENTS

Special thanks to Søren Primdahl (Topsoe Fuel Cell A/S) for academic input and discussion of the results in this work, Per Hjalmarsson (Risoe DTU) for provision of LCN60/CGO and to Alfred Samson (Risoe DTU) for provision of CGO backbones and assistance in the experimental work carried out. Thanks to Carsten Gynther (Risoe DTU) for assistance with TG measurements.

REFERENCES

[1]Jung, W., Harry L. Tuller. *Solid State Ionics,* **180**, 843-847 (2009).
[2]Rothschild, A., W. Menesklou, H. L. Tuller, E. Ivers-Tiffée. *Chem. Mater.,* **18**, 3651-3659 (2006).
[3]Patrakeev, M. V., I. A. Leonidov, V. L. Kozhevnikov, V. V. Kharton. *Solid State Sc.,* **6**, 907-913 (2004).
[4]Ullmann, H., N. Trofimenko, F. Tietz, D. Stöver, A. Ahmad-Khanlou. *Solid State Ionics,* **138**, 79-90 (2000).
[5]Brixner, L. H. *Mat. Res. Bull.,* **3**, 299-308 (1968).
[6]Kharton, V. V., et al. *Solid State Ionics,* **133**, 57-65 (2000).
[7]*SurfaceNet.* http://www.surfacenet.de/html/strontium_titanate.html (06. December 2011).
[8]Singhal, S. C., K. Kendall. High Temperature Solid Oxide Fuel Cells: Fundamentals, Design and Applications, chapter 4 & 5. *Elsevier Advanced Technology*, Oxford, 2003.
[9]Samson, A., M. Søgaard, R. Knippe, and N. Bonanos. *J. Electrochem. Soc.,* **158 (6)**, 1-10 (2011).
[10]Nicholas, J. D., S. A. Barnett. *J. Electrochem. Soc.,* **157 (4)**, B536-B541 (2010).
[11]Hjalmarsson, P., M. Mogensen. *J. Power Sources,* **196**, 7237–7244 (2011).
[12]Gharbage, B., F. Mandier, H. Lauret, C. Roux, T. Pagnier. *Solid State Ionics,* **82**, 85-94 (1995).
[13]Wang, H. B., J. F. Gao, D. K. Peng, G. Y. Meng. *Mat. Chem. Phys.,* **72**, 297-300 (2001).
[14]Takeda, Y., et al. *J. Solid State Chem.,* **63**, 237-249 (1986).
[15]Majid, A., J. Tunney, S. Argue, M. Post. *J. Sol-Gel Sc. and Tech.,* **32**, 1-3, 323-326 (2004).
[16]Iwahara, H., T. Esaka, H. Uchida, N. Maeda. *Solid State Ionics,* **3/4**, 359-363 (1981).
[17]Nag, A., T. R. N. Kutty. *Royal Society of Chemistry.*
   http://www.rsc.org/suppdata/jm/b2/b207756f/b207756f.pdf (11. December 2011).
[18]Huang, Y., J. M. Vohs, R. J. Gorte. *Electrochem. and Solid-State Let.,* **9 (5)**, A237-A240 (2006).
[19]Bidrawn, F., S. Lee, J. M. Vohs, R. J. Gorte. *J. Electrochem. Soc.,* **155 (7)**, B660-B665 (2008).
[20]Steele, B. C. H. *Solid State Ionics,* **158 (6)**, 95-110 (2000).

FREEZE-TAPE CASTING FOR THE DESIGN OF ANODE-DELIVERY LAYER IN SOLID OXIDE FUEL CELLS

Jacob Bunch[1], Yu Chen[1], Fanglin Chen[1*], and Matthew May[2]

[1] Department of Mechanical Engineering, University of South Carolina, Columbia, SC 29208
[2] Department of Physics, The Citadel, Charleston, SC 29409
[*] Author to whom correspondence should be addressed:
E-mail address: chenfa@cec.sc.edu

ABSTRACT

Solid oxide fuel cells (SOFCs) have tremendous commercial potential due to their high efficiency, high energy density, and fuel flexibility, operating on both hydrogen and hydrocarbon fuels. The material and fabrication challenges of SOFCs have been extensively studied and well documented. The design for this type of fuel cell is critical if this technology will become widely used and accepted. The method proposed in this study is a novel technique known as freeze-tape casting. This method is used to prepare the anode by tailoring to desired porosity. The work discussed in this paper characterizes the freeze-tape casting method on the final microstructure and performance of the nickel-yttria-stabilized zirconia (Ni-YSZ) anode. In fuel cell mode, the cell demonstrates low polarization resistance of 0.16 Ω cm^2 at 800°C with H$_2$ as fuel and ambient air as oxidant with Ni-YSZ as the anode, YSZ as the electrolyte, and (La$_{0.75}$Sr$_{0.25}$)$_{0.95}$MnO$_3$ (LSM)-YSZ as the cathode. The objective of this work is to understand the roles this method has on the microstructure of the anode and performance of the cell. The knowledge will be then used with the intention of optimizing the SOFC anode.

1. INTRODUCTION

Solid oxide fuel cells (SOFCs) have attracted great interest due to their high efficiency, high energy density, and fuel flexibility. A SOFC typically consists of three components: a fuel electrode (anode), an air or oxygen electrode (cathode), and an electrolyte which can be seen in Figure 1. SOFCs typically have two basic designs: electrolyte supported cells and electrode supported cells. Electrolyte supported cells typically have high cell ohmic resistance resulted from the relatively thick electrolyte which normally has low electrical conductivity. For this reason, electrode supported cell was selected as the cell design for this study. The success of the SOFC is heavily dependent on the design and development of this technology. Material selection is critical to the performance of the cell. Among the possible choices of the anode material, Ni–YSZ has been widely used due to the high electrical conductivity provided by the metallic Ni and the good ionic conductivity afforded by the YSZ.[1]

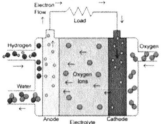

Figure 1. Schematic diagram of the operational principles of an SOFC

For a SOFC, the anode should be porous, which allows the fuel to travel toward reaction sites for oxidation. In general, the reaction is limited by the length of the so called triple-phase boundary

(TPB).[1] TPB is the area at which the ionic conducting phase, fuel gas, and the electronic connecting catalyst contact on each other. In our case, this is where the Ni anode catalyst, the YSZ electrolyte, and the fuel gas are in contact with each other. A simplified schematic of the TPB is shown in Figure 2.

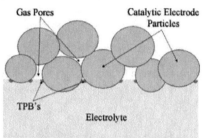

Figure 2: A simplified schematic diagram of the anode/electrolyte interface which illustrates the TPB reaction sites.

The optimization of the TPB is critical for the improved performance of SOFCs. It is extensively reported that modification to the electrode structure is able to minimize both activation and concentration polarizations, while having reported power densities greater than 1 Wcm^{-2} at 600°C with a highly porous die-pressed nickel cermet anode supported solid oxide fuel cell by controlling the microstructure of the anode.[2; 3; 4] A method that currently aims to improve the TPB is freeze-tape casting.[5] Freeze-tape-casting, which combines tape-casting and freeze-casting processes both of which have been applied in the field of manufacturing catalyst support structures, solid oxide fuel cells and filtration membranes[6; 7], embodies the ability to produce a thin planar substrate capable of forming and controlling complex pore structure. Directionally graded acicular pores (<5-100) μm) are very helpful for gas delivery in the electrode and can be achieved through freeze-tape casting. However, freeze-tape casting forms macro pores in both the bottom and top of the anode substrates, drastically decreasing the ability to apply an effective thin electrolyte layer. A thin electrode active layer has proven to be an effective method of overcoming this restriction to improve the adhesion between the electrode and electrolyte.[8; 9]

Freeze-tape casting holds advantages over traditional methods in that columnar pore morphology and resultant forces may also promote extension of the TPB region.[10] The surface morphology of freeze-tape cast anodes permits significant electrolyte coating penetration, thus extending the critical TPB region, while also minimizing anode/electrolyte degradation induced by delamination.[10]

Traditional methods of anode preparation all have advantages and disadvantages. Traditional methods of anode-supported SOFC preparation include but are not limited to electrochemical vapor deposition (EVD), chemical vapor deposition (CVD), infiltration method, screen-printing, and tape casting.[11] For example, the impregnation of nano-sized Ni into the YSZ backbone can greatly improve the cell performance, but has a disadvantage when it comes to the coarsening of Ni. This can occur when the Ni nanoparticles are exposed to high temperatures causing these particles to grow in size. This growth causes the Ni within the YSZ backbone to agglomerate together further causing performance degradation. Freeze-tape casting avoids this by forming acicular pore within the Ni-YSZ anode support layer. In anode support layer of Ni-YSZ, the YSZ acts as an inhibitor for the coarsening of Ni phase during consolidation and operation. The current work is focused on characterizing the novel microstructure of the anode substrate fabricated by the freeze-tape casting method as well as studying the cell performance.

2. EXPERIMENTAL PROCEDURE

2.1 Equipment Design

The freeze-tape casting apparatus was designed from the specifications within the literature.[12] The freeze-tape casting equipment consists of a standard research tape caster (TTL-1200; Richard Mistler Inc.). The drying bed has been adapted with cooling lines and a low-temperature recirculating chiller (Neslab ULT-80; Thermo Electron Corp., Waltham, MA). To ensure efficient cooling and uniform temperature distribution, heat transfer plates (Thermofin C; Radiant Engineering Inc.) and thermal epoxy (50-3185R; Epoxies Etc..., Cranston, RI) were used along with foam insulation to thermally isolated from the rest of the unit to introduce a sharp temperature transition. This design allows the cast slurry to flow and level before the solidification process while allowing control of the freezing platen temperature to modify the tape properties.[12] The freeze-tape casting apparatus can be seen below in Figure 3.

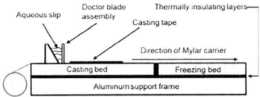

Figure 3. Modified freeze-tape casting apparatus.[12]

2.2 Experimental Design

The experiment was designed by following the flowchart in the literature.[12] Nickel oxide (NiO) (J.T. Baker) and yittria-stabilized zirconia (YSZ) doped with 8 mol% Y_2O_3 (TZ-8Y; Tosoh Co., Tokyo, Japan) were used as the starting powders. A powder ratio of 6 NiO to 4 YSZ was kept constant throughout the experiment. The powders were then prepared as aqueous suspensions with an ammonium polyacrylate dispersant (Darvan C-N, R.T. Vanderbilt Co., Inc., Norwalk, CT) and an acrylic latex emulsion binder system (Duramax, Rohm & Haas, Philadelphia, PA). These mixtures were heated for up to 3 hours, mixing every hour, to ensure all components of the suspension were homogenous.

Once the desired mixture is achieved, the slurry is then cast in to approximately 4 in. x 12 in. tapes and allowed to solidify on the freezing bed at a constant temperature. A constant pulling rate of < 10 mm/s was utilized to ensure a continuous and directional solidification front over the length of the tape, which is essential in the formation of long-range pore ordering.[12] After solidification, the samples were cut into 12 mm diameter circular disks and placed in a freeze dryer (VirTis AdVantage 2.0 BenchTop Freeze Dryer, SP Scientific, Stone Ridge, NY) at a constant shelf temperature of -30°C at 80 mTorr. The condenser temperature is kept at a constant temperature of -50°C to ensure sublimation occurs. The samples were kept in the freeze dryer for up to 24 hours before sintering at 1100°C for 5 hours with a heating rate of 1°C/min. A special binder burnout process is not needed due to the thin ceramic substrate and the porosity present in the green state which improves the polymer decomposition gases to escape.[12]

Aqueous Ni-YSZ slurries were prepared at 20 vol% solids loading and frozen at -40°C to characterize the microstructure and performance. The NiO-YSZ active layer and YSZ electrolyte were prepared by airbrushing each on to the NiO-YSZ delivery layer. Non-aqueous NiO-YSZ and YSZ slurries were prepared for airbrushing. The samples were co-sintered at 1400°C for 2 hours with a heating rate of 1°C/min. The cathode was prepared using screen printing method of an LSM-YSZ ink. The samples were then sintered at 1100°C for 2 hours with a heating rate of 1°C/min.

2.2 Characterization

The microstructure of the substrate was characterized using a scanning electron microscope (SEM, FEI Quanta 200). The current density-voltage curves as well as the impedance spectra were measured with a four-probe method using a multi-channel Versa STAT (Princeton Applied Research) at the operating temperature range from 700°C to 800°C. The high temperature fuel cell and electrolysis testing system can be seen in our previous work.[13] During fuel cell testing, humidified hydrogen gas (3 % H_2O/97% H_2) was used as fuel to the anode side while the cathode side was exposed to the ambient environment. The cell polarization resistance (R_p) was determined from the difference between the low and high frequency intercepts of the impedance spectra with the real axis.

3. RESULTS AND DISCUSSION

3.1 Novel Microstructure Characterization

During the freeze-tape casting, micron-sized ice crystals initiate at the Mylar™ carrier side causing larger and larger ice crystals to grow toward the topside of the green tape. Figure 4 demonstrates a simplified cross-section of the solidification process of an aqueous system. During the solidification, the Ni-YSZ slurry changes into ice crystals and ceramic walls due to the ceramic particles being rejected by the growing ice crystals.

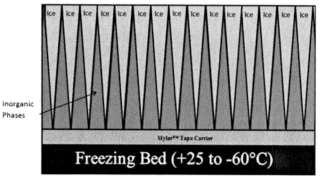

Figure 4. Solidification schematic of aqueous solvent system.[12]

Once the green tape has been further processed through freeze drying, a natural gradient of porosity is formed. Figure 5 shows a Ni-YSZ substrate after sublimation, but before sintering. It can be easily seen that the finger-like acicular pores have grown perpendicular to the freeze bed. Figure 6 shows the same sample after sintering. The porosity of the substrate after sintering increases a small amount due to combustion of the organic materials within the green sample. To further validate Figure 4, SEM image were taken of the Mylar side (bottom) and the atmospheric side (top) of the sample. The bottom side, as demonstrated in Figure 4 and seen in Figure 7a, has very small pore sizes that range from about 1-2µm. For this reason, the bottom side is used as the base for the active layer and electrolyte. The top side, which can be seen in Figure 7b, has much larger pores that range from about 20-30µm. The top side can serve as the gas delivery layer.

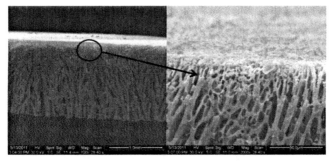

Figure 5. Green sample before sintering.

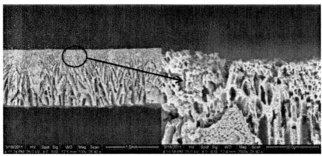

Figure 6. Sample after sintering.

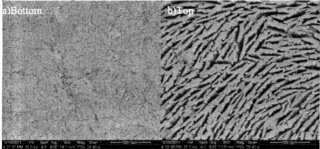

Figure 7. Surface morphology. a) bottom surface. b) top surface.

Figure 8 shows the interfaces of the novel microstructure of an anode-supported SOFC, the Ni-YSZ functional layer, as well as the dense YSZ electrolyte. The cell has thick porous NiO-YSZ delivery layer, thin porous NiO-YSZ functional layer, dense YSZ electrolyte layer, and porous LSM-YSZ layer with thickness of about 1000, 20, 20, 30 μm, respectively. Also, electrolyte coating penetration into the functional layer can be seen in Figure 8b. This is necessary to ensure optimization of the TPB and good adhesion between the two layers.

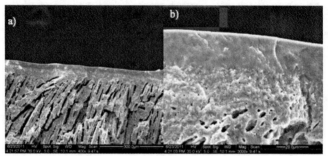

Figure 8. Delivery layer, functional layer, electrolyte interface. a) tri-layer interfaces b) active-layer and electrolyte interface

From Figures 6 and 8, it can be seen that novel acicular and graded pores have been fabricated by employing the freeze-tape casting method. The existence of acicular pores can improve fuel delivery, fuel mixing, and enable better gas diffusion to the TPB in the functional layer accelerating the reactions. The obtained dense defect-free YSZ is well adhered to the electrodes and should provide good ionic conduction.

3.2 Performance Characterization

Electrochemical performance evaluation was performed on cells shown in Figure 8 under the SOFC mode at different temperatures. Figure 9 shows the cell voltage, current and power output curves. The cell peak power densities are 0.16, 0.22, 0.25 $W \cdot cm^{-2}$ at 700, 750, and 800 °C, respectively, which is much less than the values previously reported at the same testing conditions with the similar cell materials and cell configuration.[14]

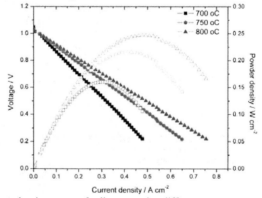

Figure 9. V-I and I-Power density curves of cell measured at different temperatures.

However, the cell impedance spectrum displays a different picture. The cell ohmic resistance (R_Ω), which corresponds to the high-frequency intercept of the impedance spectra with the real axis in the Nyquist plot, is shown in Figure 10. The cell polarization resistance (R_P), determined from the difference between the high- and low- frequency intercepts of the impedance spectra with the real axis, is about 0.56, 0.29, and 0.16 Ω cm^2 at 700, 750 and 800°C, respectively. As shown in Figure 11, in comparison to other publications[15; 16; 17], the total electrode polarization resistances are much lower. The reduced electrode polarization may be resulted from the improved gas diffusion and extended reaction area in the anode. The inferior cell performance can be attributed to a lower OCV and higher ohmic resistance. In comparison with the reported values by Y. Leng[14] and L. Zhang[16], 1.12 V and 1.11V at 800°C, the cell's OCV of 1.03V at 800°C is much lower. The ohmic resistance reported by both Y. Leng[14] and L. Zhang[16] are around 0.2 Ω cm^2 , while the cell used in this experiment was around 0.7 Ω cm^2. The low OCV and high ohmic resistance can be attributed to the defects within the electrolyte layer.

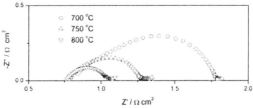

Figure 10. Impedance spectra measured at different temperatures under open circuit conditions.

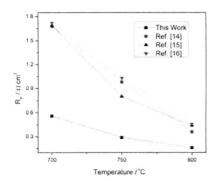

Figure 11. Comparison of R$_p$ with previous publications.

4. CONCLUSION

The novel freeze-tape casting method was used to prepare functional graded and acicular anode support for SOFCs. The microstructure of the cells with the novel finger-like acicular pores provides an excellent support and delivery layer for the anode supported SOFCs. The novel structure has shown an impressive impact on the electrode polarization resistance. The cell demonstrates polarization resistances (R$_p$) of 0.56, 0.29, and 0.16 Ω cm^2 at 700, 750 and 800°C, respectively. The results from the polarization study indicate that the novel functional graded acicular electrode has greatly reduced the resistance to gas mixing and diffusion, thus enhancing the cell polarization performance. From these results, additional research can be conducted to further understand this method in hopes of the optimization of the anode delivery-layer and further cell performance enhancement.

5. FUTURE WORK

Future work will consist of microstructure control and characterization. Microstructure control is very important in the optimization of the freeze-tape casting method. Production variables such as solids loading, freezing temperature, and pulling rate will be focused on initially. Other variables including but not limited to viscosity, binder wt%, and dispersant wt% may also be taken into account to understand the effects on the anode-support layer.

ACKNOWLEDGEMENT

Financial supports from the NASA EPSCoR (Award Number NNX10AN33A) and the South Carolina Space Grant Consortium are gratefully acknowledged.

REFERENCES

[1]Y.-H. Koh, J.-J. Sun, andH.-E. Kim, "Freeze casting of porous Ni–YSZ cermets," *Materials Letters,* **61**[6] 1283-87 (2007).

[2]T. Suzuki, Z. Hasan, Y. Funahashi, T. Yamaguchi, Y. Fujishiro, andM. Awano, "Impact of Anode Microstructure on Solid Oxide Fuel Cells," *Science,* **325**[5942] 852-55 (2009).

[3]M. H. D. Othman, N. Droushiotis, Z. T. Wu, G. Kelsall, andK. Li, "High-Performance, Anode-Supported, Microtubular SOFC Prepared from Single-Step-Fabricated, Dual-Layer Hollow Fibers," *Adv Mater,* **23**[21] 2480-+ (2011).

[4]J. W. Kim, A. V. Virkar, K. Z. Fung, K. Mehta, andS. C. Singhal, "Polarization effects in intermediate temperature, anode-supported solid oxide fuel cells," *J Electrochem Soc,* **146**[1] 69-78 (1999).

[5]"Fabrication of Functionally Graded and Aligned Pores Via Tape Cast Processing ", *NASA Tech Brief,* **LEW-17628-1** (2007).

[6]M. Nagamori, T. Shimonosono, S. Sameshima, Y. Hirata, N. Matsunaga, andY. Sakka, "Densification and Cell Performance of Gadolinium-Doped Ceria (GDC) Electrolyte/NiO-GDC Anode Laminates," *Journal of the American Ceramic Society,* **92**[1] S117-S21 (2009).

[7]K. Lindqvist and E. Lidén, "Preparation of alumina membranes by tape casting and dip coating," *J Eur Ceram Soc,* **17**[2-3] 359-66 (1997).

[8]M. D. Gross, J. M. Vohs, andR. J. Gorte, "A strategy for achieving high performance with SOFC ceramic anodes," *Electrochem Solid St,* **10**[4] B65-B69 (2007).

[9]Z. Wang, N. Zhang, J. Qiao, K. Sun, andP. Xu, "Improved SOFC performance with continuously graded anode functional layer," *Electrochem Commun,* **11**[6] 1120-23 (2009).

[10]P. Gannon, S. Sofie, M. Deibert, R. Smith, andV. Gorokhovsky, "Thin film YSZ coatings on functionally graded freeze cast NiO/YSZ SOFC anode supports," *Journal of Applied Electrochemistry,* **39**[4] 497-502 (2009).

[11]C. M. An, J. H. Song, I. Kang, andN. Sammes, "The effect of porosity gradient in a Nickel/Yttria Stabilized Zirconia anode for an anode-supported planar solid oxide fuel cell," *J Power Sources,* **195**[3] 821-24 (2010).

[12]S. W. Sofie, "Fabrication of Functionally Graded and Aligned Porosity in Thin Ceramic Substrates With the Novel Freeze?Tape-Casting Process," *Journal of the American Ceramic Society,* **90**[7] 2024-31 (2007).

[13]C. H. Yang, A. Coffin, andF. L. Chen, "High temperature solid oxide electrolysis cell employing porous structured (La(0.75)Sr(0.25))(0.95)MnO(3) with enhanced oxygen electrode performance," *Int J Hydrogen Energ,* **35**[8] 3221-26 (2010).

[14]Y. Leng, "Performance evaluation of anode-supported solid oxide fuel cells with thin film YSZ electrolyte," *Int J Hydrogen Energ,* **29**[10] 1025-33 (2004).

[15]Y. Chen, F. Chen, W. Wang, D. Ding, andJ. Gao, "Sm0.2(Ce1−xTix)0.8O1.9 modified Ni–yttria-stabilized zirconia anode for direct methane fuel cell," *J Power Sources,* **196**[11] 4987-91 (2011).

[16]L. Zhang, S. P. Jiang, W. Wang, andY. J. Zhang, "NiO/YSZ, anode-supported, thin-electrolyte, solid oxide fuel cells fabricated by gel casting," *J Power Sources,* **170**[1] 55-60 (2007).

[17]L. Zhang, J. Gao, M. Liu, andC. Xia, "Effect of impregnation of Sm-doped CeO2 in NiO/YSZ anode substrate prepared by gelcasting for tubular solid oxide fuel cell," *J Alloy Compd,* **482**[1-2] 168-72 (2009).

TAILORING THE ANODE MICROSTRUCTURE IN MICRO-TUBULAR SOFCS BY THE OPTIMIZATION OF THE SLURRY

Michele Casarin, Riccardo Ceccato and Vincenzo M. Sglavo,
Department of Materials Engineering and Industrial Technologies,
University of Trento, Trento, Italy.

ABSTRACT

Tailoring the anode microstructure is fundamental for the improvement of solid oxide fuel cell performance especially for anode-supported devices. In the present study, the improvement of suspension stability is achieved by the optimization of dispersant with the effect of more stable suspensions over time, minimizing the effect of external conditions such as humidity and temperature, for a better control and reliability of wet colloidal process. Moreover, a good dispersion of suspension allows to obtain homogeneous microstructure in fired bodies derived from highly concentrated slurries with solid load in excess to 80 wt%. In addition, high packing efficiency of high concentrated suspensions as well as a pore former characterized by anisotropic shape are responsible for lowering the percolation limit of porosity. In this condition a percolating pore network can be achieved at porosity value as low as 13 vol% in fired sample.

INTRODUCTION

Recently, solid oxide fuel cell (SOFC) technology has gained much attention as potential energy source and many efforts have been made to commercialize both planar and tubular SOFCs.[1] Tubular and especially micro-tubular solid oxide fuel cells (MT-SOFCs) have remarkable advantages in comparison with planar SOFCs such as smaller sealing area, less problematic gas feeding and higher thermal resistance.[2] Additionally MT-SOFCs offer higher power density and higher thermal shock resistance as the cell diameter decreases.[3] Nevertheless as the cell diameter is scaled down the current collection represents a critical issue with regard to the cell performance. In fact, as the cell size diminishes the contact between electrode and current collector becomes difficult. It was proved that cell resistance decreases with the increase of the contact area between the electrode and the current collector.[4] The production of current collector-supported MT-SOFCs in which a nickel wire is embedded within the anode substrate is an effective approach to improve the cell performance by decreasing the cell resistance.[5] This current collector-supported MT-SOFC is produced by dip-coating process and consists of NiO/YSZ anode with embedded current collector, YSZ electrolyte and a two-layered cathode of YSZ/LSM and pure LSM. Studies have shown that a further increase of cell performance can be obtained by producing a two-layered anode.[6] The multilayered anode is composed of a gas diffusion layer and a functional layer close to the electrolyte: the former enhances the transport of fuel gases and the removal of product gases and therefore the concentration polarization is decreased. The purpose of the functional layer is to maximize the Ni/YSZ/gas reaction area at triple phase boundaries (TPBs). The effect is an increase of electrochemical reaction sites thus lowering the activation polarization.[7] One of the fundamental issues associated to the production of the anode in MT-SOFC by dip coating is the viscosity control of the suspension and the realization of a structure with interconnected porosity starting from high concentrated suspensions that allow a limited shrinkage upon drying and sintering. The main purpose of the present work is the improvement of anode suspension stability for a better control and reliability of wet colloidal process, thus minimizing the influence of external conditions such as humidity and temperature. It is recognized that temperature has a strong effect on the rheology especially for suspensions of high solid content.[8,9] Moreover, it was shown that the temperature plays an important role on the dispersant adsorption and stability of aqueous suspensions and rheological measurements were adopted to investigate the dispersion state of the suspensions.[10-12] The second and concomitant objective of the work is to tailor the anode microstructure through an optimal tuning of colloidal suspensions for the anode multilayer fabrication. The dispersant controls the suspension properties such as viscosity:[11] the lower is the viscosity, the better the particles are dispersed that means a better particle packing efficiency and homogeneous green body microstructure.[14] Moreover, a poor powder dispersion indicates the presence of powder agglomerates and causes an increase of the suspension yield stress.[13] Poor powder dispersion is ascribed to lower coverage of dispersant onto particle surface and it is related to an inadequate dispersant amount.[14] In addition, the yield stress is also associated to the force acting between particles and the liquid medium: the greater is the yield stress, the greater is the attractive interparticle potential and the stronger is the particle network.[13] The final aim of the present work is the adjustment of dispersant amount in order to lower the suspension viscosity, improve the suspension stability and point out the effect of solid loading on fired microstructure.

EXPERIMENTAL PROCEDURES

Suspension Preparation

NiO (J.T. Baker Inc., USA) and 8 mol% Y_2O_3-ZrO_2 (TZ-8YS, Tosoh, Japan) oxide powders were used for the anode substrate and the porosity was increased by adding graphite powder (Flake, 7-10 μm, Alfa Aesar, Germany) as pore former. The different powders and dispersant (Darvan 821A, Vanderbilt, Norwalk, CT) were mixed to deionized water for 2 h in a high energy rotatory mill (Turbula T2F, Bachofen, Switzerland) as reported in previous works.[5] The pore former covers with 5 wt% the total solid loading whereas the remaining solid content interests NiO and YSZ in 58:42 weight ratio.

Screening Suspensions

The effect of dispersant interests screening suspensions characterized by 70 wt% solid loading. The dispersant is added on dry powder base and its optimum value was selected to correspond to the lowest viscosity value evaluated at 0.85 s^{-1} shear rate. The concentration and taxonomy of the suspensions are reported in Table I. Rheological measurements were performed by hysteresis technique consisting in increasing and decreasing shear ramps between zero and a maximum value.[15] This method allows to investigate time effects such as thixotropy and involves an ascending ramp from 0.85 to 70 s^{-1} a dwell at maximum shear rate for 50 s followed by a descending ramp from 70 to 0.85 s^{-1}. All measurements were carried out at 23°C using rheometer (Anton Paar, MCR-301) with cone and plate system in controlled shear rate mode and the estimation of yield stress was evaluated by applying the Casson expression.[18] In addition, the suspension stability over time was determined by monitoring the solvent evaporation under laboratory conditions of relative humidity and temperature. The surface area of evaporation and stirring velocity were maintained constant in order to supply qualitative information of the dispersant effect on water evaporation.

Table I. Dispersant concentration and taxonomy of screening suspensions.

Dispersant concentration [dpb%]	0.615	0.63	0.69	0.76	0.91	0.96	1.02	1.04	1.20
Suspension taxonomy	pristine	+2.3	+12	+23	+45	+57	+66	+69	+95

High Solid Content Suspensions

Suspensions with high solid loading were prepared with optimum dispersant concentration, i.e. 1.04 dpb% and the solid loading of 80, 83 and 85 wt%. These suspensions and corresponding fired samples are labelled here as 80D, 83D and 85D respectively. Rheological analysis was carried out using the same conditions described before. The effect of solid loading was evaluated by measuring the viscosity at shear value of 0.85 s^{-1} as well as the dependence of viscosity on shear rate for 80D, 83D and 85D suspensions with the comparison of 70 wt% slurry. The flow behavior of suspensions was characterized by the yield stress, power law index and Krieger-Dougherty model.[9] The maximum solid concentration is obtained by using Doroszkowski method and the flow resistance parameter is then calculated as reported in other studies.[14,18] Dilatometry analysis was carried out on samples prepared from 80D, 83D and 85D suspensions by using a horizontal dilatometer (Linseis L75). These samples were sintered at 1380°C for 3 h at heating rate of 5°C/min and organics and pore former were removed with dwell steps at 400 and 800°C for 1 h. Finally, fired samples were observed by SEM and the open porosity was measured via Archimede's method and averaged on three samples measurement. In addition, an estimation of porosity left by the burnout of pore former was estimated by using Slamovich-Lange expression.[21]

RESULTS & DISCUSSION

Screening Suspensions
    The effect of dispersant on the suspension viscosity is shown in Figure 1 (a). The lowest viscosity value is obtained at dispersant concentration of 1.04 dpb%. In this condition, the surface coverage of dispersant onto particle surface is maximized and corresponds to the lower extent of powder agglomeration or flocculation.[14] Experimental data are well fitted by a quadratic regression model. In this way the estimated viscosity at pristine value of dispersant is 5.92 Pas. For the optimum dispersant concentration of 1.04 dpb% the viscosity value is 1.17 Pas that is five times lower than the estimated viscosity at pristine condition, i.e. 0.615 dpb% of dispersant, for 70 wt% suspension evaluated at 0.85 s^{-1} shear rate at 23°C. The diagram in Figure 1 (b) points out the

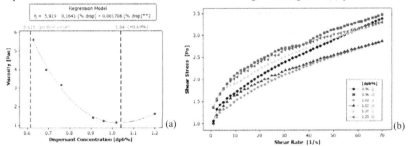

Figure 1. Viscosity vs dispersant concentration for 70 wt% suspensions (a) and shear stress vs shear rate (i) ascending ramp, ii) descending ramp) for suspensions prepared with 0.96, 1.02 and 1.20 dpb% of dispersant respectively +57, +66 and +95 suspensions (b).

lowest viscosity along with the lowest hysteresis loop for the slurry with dispersant amount close to the optimum value, i.e. 1.02 dpb%. In this case the ascending and descending ramps are almost superimposed and this means that the microstructure of suspension is far not affected by the shear history and stable over the time involved for the measurement. For +95 and especially for +57 suspension the hysteresis loops are quite evident and the suspensions microstructure depends on shearing. For both suspensions, powder agglomeration occurs along the descending ramp and viscosity is increased. Hydrodynamic forces at high shear rates induce the formation of agglomerates and Brownian motion is not able to restore the previous microstructure during the experiment time.[16] On the other hand irreversible changes of suspension microstructure can be related to evaporation of suspension solvent. Under this assumption, the suspension stability over time was also evaluated by monitoring the solvent evaporation under laboratory conditions of relative humidity and temperature. From Figure 2 (a) one can see that the lowest amount of water and water loss coefficient (m) correspond again to +66 suspension. In this case the relative humidity

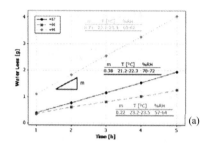
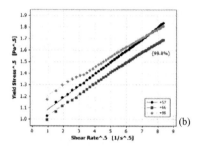

Figure 2. Water loss function vs time where m is the water loss coefficient and %RH the relative humidity (a) and yield stress$^{0.5}$ vs shear rate$^{0.5}$ calculated by Casson model with correlation coefficient of 99.8% (b) for +57, +66 and +95 suspensions.

is in the range of 57-64% that is similar to that regarding +95 suspension, (60-62 %RH), which displayed a much higher water loss in spite of lower temperature. Moreover, the lowest evaporation should occur at lower temperature and higher RH as for the conditions of +57 suspensions, i.e. 21.3-22.3°C and 70-72 %RH. As a matter of fact, this does not occur and consequently the dispersant concentration should be expected to affect the evaporation of water. In addition a linear dependence of water loss over the time is evident within 57-72 RH% and 21.2-23.5°C ranges for the +57, +66 and +95 suspensions. This linear trend is similar to that regarding the rate of evaporation during constant rate period (CRP) involved in ceramic materials drying.[17] Further insights on the effect of dispersant on suspension stability may arise from the evaluation of solvent evaporation under controlled laboratory conditions of relative humidity and temperature adopting the equation:[17]

$$\dot{V}_E = H\,(P_W - P_A) \tag{1}$$

where $\dot{V}_E$ is the evaporation rate, P_W the vapor pressure of the liquid at the surface and P_A the ambient vapor pressure. H is a factor depending on the surface area of evaporation and other conditions such as stirring. In the current investigation the surface area of evaporation and stirring velocity were maintained constant and consequently the water loss by the suspensions should be related to the difference of vapor pressure terms in the Eq. (1) to the temperature and relative humidity. All suspensions in Figure 1 (b) display shear thinning behavior and the yield stress is estimated by Casson expression. The yield stress τ_y is calculated by extrapolating at zero shear rate the Casson expression:[18]

$$\tau^{1/2} = \tau_y^{1/2} + c\gamma^{1/2} \tag{2}$$

The lowest water loss and hysteresis area as well as the lowest flow resistance parameter and yield stress occur for +66 suspension with dispersant close to the optimum value, i.e. 1.02 dpb% as shown in Table II and Figure 2 (b). The optimum dispersant amount adsorbed onto particle surface is responsible for a good dispersion state of suspension that means a low powder agglomeration and therefore a low yield stress. In conclusion, the suspension with the optimum dispersant concentration shows a lower viscosity as well a more stable microstructure along time in terms of hystereris and water loss due to the full coverage of particle surface by the dispersant that correspond to the minimum degree of agglomeration.[14] A better stability of colloidal suspensions with optimum dispersant concentration improves the control and the reliability of wet colloidal process thus minimizing the influence of external conditions as humidity and temperature and a more homogeneous microstructure of the green body can also be expected.[14,23]

Table II. Screening suspension properties: yield stress, maximum solid concentration (A), relative degree of freedom of dispersed particles (A-ø), flow resistance parameter (ø/(A-ø)) and hysteresis area.

Dispersant concentration [dpb%]	Yield stress [Pa]	A	A-ø	ø/(A-ø)	Hysteresis area
0.96	.97	.380	.086	3.42	19
1.02	.89	.403	.109	2.70	4.5
1.20	1.27	.399	.105	2.80	6.2

High Solid Content Suspensions

Powder concentration has a great influence on rheological behavior especially for high solid loading suspensions. The viscosity increases linearly from 70 to 83 wt% suspension as shown in Figure 3 (a). For the 85D slurry the viscosity increases greatly and agglomeration probably takes place. Figure 3 (b) shows a shear thinning behavior for 70, 80, 83D suspensions and the viscosity increases with the solid loading as expected.[9] The 85D suspension moves partially away from a shear thinning behavior. In fact the 85D viscosity curve tends to lay to a constant value at higher shear rates. This trend resembles the behavior of non-Newtonian fluids characterized by two plateau of constant viscosity, Newtonian regions, which occurs at very low and very high shear rates.[8] It is evident that beyond a certain solid concentration, i.e. 83 wt%, repulsive forces between particles are not effective anymore regardless of a good surface coverage by the dispersant. This occurs for 85D slurry where hydrodynamic forces prevail over repulsive forces at high shear rates and the attractive particle-particle interactions are favored by the reduced interparticles distance occurring at high solid concentrations.[9] For 70, 80 and 83D suspensions the shear thinning effect is evident and the

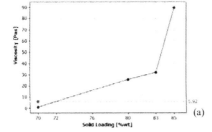
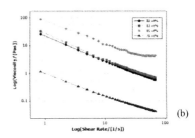

Figure 3. Viscosity vs solid loading (a) and dependence of viscosity on the shear rate (b) for 70, 80, 83 and 85 wt% suspensions.

viscosity reduces with the shear rate because of shear induced disruption of the agglomerates.[24] Conversely, for 85D system probably an equilibrium between formation and disruption of agglomerates is achieved at higher shear rates.[26] The optimization of dispersant allows reaching higher solid loading due to a better particle packing efficiency. For instance as shown in Figure 3 (a) at viscosity of 5.92 Pas corresponds a 72 wt% solid loading for the slurry with optimized dispersant but also a 70 wt% solid loading suspension without optimized dispersant as pointed out in Figure 1 (a) for pristine suspension (0.615 dpb%). The optimization of dispersant allows to obtain higher solid suspensions for a given viscosity value. In addition the flow behavior of suspensions can be described by an empirical power-law dependence:[22]

$$\eta = K\gamma^{n-1} \qquad (3)$$

where η is the viscosity, K the consistency index and n is the power-law index or shear thinning constant. n values lower than 1 indicate a departure from Newtonian behaviour towards an

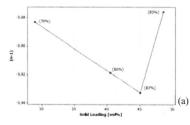
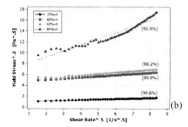

Figure 4. Power law index (n-1) vs solid loading (a) and Casson model (correlation coefficient in parenthesis) (b) for 70, 80, 83 and 85 wt% suspensions.

increasing shear thinning behavior. In Figure 4 (a) and Table III it is evident that the index n decreases with solid loading and consequently an increasing of shear thinning behavior occurs from 70 to 83 wt% of solid loading. Beyond this solid concentration the trend changes to a greater n index value and another flow behaviour should be expected. The power law model confirms the closer Newtonian trend of flow curve regarding 85D system as shown in Figure 3 (b). Once again the 85D slurry is not well fitted by Casson model as a further confirmation of a weak shear thinning behavior compared to the 70, 80 and 83D suspensions which are well fitted by

Table III. High concentrated suspensions: yield stress and power law index (n), maximum solid concentration (A), volume particle fraction (ø) and flow resistance parameter (ø/(A-ø)).

Suspension [%wt]	Yield stress [Pa]	n	(n-1)	A	ø	ø/(A-ø)
70	0.89	.192	-.808	.3980	.2899	2.68
80	20.7	.103	-.897	.4395	.4048	11.8
83	23.5	.068	-.932	.4867	.4508	12.6
85	50.6	.210	-.790	.4962	.4857	46.3

Casson model as shown in Figure 4 (b). For all suspensions the yield stress increases with the solid loading as expected.[9] The yield stress is also linearly related to the flow resistance parameter ø/(A-ø) when shear thinning suspensions, i.e. 70, 80, 83D ones, were taken in account as shown in Figure 5 (a). The increase of flow resistance parameter indicates an increase of powder agglomeration for the 70, 80 and 83 wt% suspensions. When 85D system is considered the linear behavior between yield stress and flow resistance parameter is lost and this can be associated to the shift of shear thinning region to lower shear rates as shown in Figure 3 (b). This can be explained by assuming the presence of agglomerates and/or particles larger than 1 µm.[19] The rheological behavior of slurries and especially for those highly concentrated is fitted by Krieger-Dougherty equation, here written in linear form:[9]

$$\ln(\eta_r) = -[\eta] \, A \ln\left(1 - \frac{\varnothing}{A}\right) \qquad (4)$$

where η_r is the relative viscosity, $[\eta]$ the intrinsic viscosity, A the maximum solid concentration, $\varnothing$ the volume fraction of particles including oxide powders and pore former and $\varnothing/A$ represents the relative density or packing fraction of slurry. In Figure 5 (b) the rheology of suspensions is well

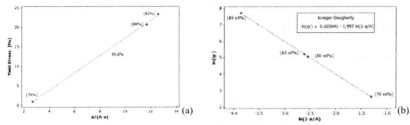

Figure 5. Yield stress vs flow resistance parameter with coefficient of correlation, c.c., of 99.8% (a) and Krieger-Dougherty model (b) with c.c. of 100.0% for 70, 80, 83 and 85 wt% suspensions.

fitted by Krieger-Dougherty model over the entire solid loading, 70-85 wt% with slope of 1.997 very close to theoretical value of 2. This enables for a given slurry viscosity to associate a precise packing fraction of slurry $\varnothing/A$ and finally a precise chemical composition in terms of $\varnothing$ and $\varnothing_{pf}$ of slurry system as evident in Table IV. As the solid loading increases the packing fraction of slurry increases up to values larger than typically 0.64 for the "jammed packing" of monosize spheres.

Table IV. Highly concentrated suspensions: volume fraction of powders ($\varnothing$), pore former fraction ($\varnothing_{pf}$), maximum solid concentration (A), intrinsic viscosity ($[\eta]$) and ratio of volume fraction of powders on maximum solid concentration ($\varnothing/A$).

Suspension [wt%]	$\varnothing$	$\varnothing_{pf}$	A	$[\eta]$	$\varnothing/A$
70	.2899	.1698	.3980	5.0	.728
80	.4048	.1513	.4395	4.5	.921
83	.4508	.1465	.4867	4.1	.926
85	.4857	.1434	.4962	4.0	.979

Asymmetric particles such graphite powder (Figure 6 (d)) can assume hexagonal order (two-dimensional) at high shear rates. In these conditions, a good dispersion of highly concentrated suspensions with rod-like particles can achieve packing fraction values close to 0.96.[25] The microstructure of fired samples obtained from 80, 83 and 85 wt% suspensions are shown in Figure 6. The uniformity of microstructure for 80 and 83D samples in terms of homogeneity and porosity are evident. In addition, agglomerates when present are smaller than those regarding 85D sample. In this latter case the microstructure is characterized by dimples, agglomerates and coarsening resulting in a poor microstructural homogeneity. Moreover, the shape of pores resembles that of graphite particles as expected after burnout and complete densification of matrix. The heat treatment response of the 80, 83 and 85D samples is shown in Figure 7 (a) and Table V. For all samples the maximum value of linear shrinkage is attained after 2 h at maximum temperature of 1380°C. The shrinkage curves are superimposed up to 1380°C and after the shrinkage values

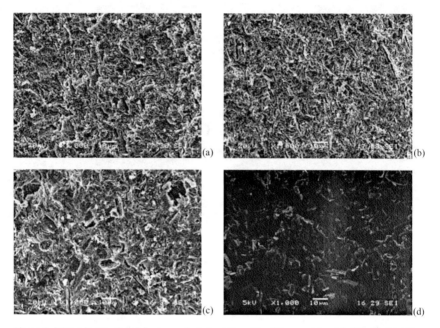

Figure 6. SEM micrographs of fired sample derived from 80 wt% suspension (a), 83 wt% (b), 85 wt% (c) and graphite particles (d).

diminish with the solid loading for 80 and 83D. The different behaviour of 85D sample can be related to the presence of agglomerates that hinders the sintering process and prevent the attaining of uniform microstructure during the last step of heat treatment. In Figure 7 (b) the porosity follows a decreasing trend with solid loading. The open porosity curve coincides with that regarding the total porosity $\emptyset_{TP}$ left by pore former burnout and calculated by Slamovich-Lange expression:[21]

$$\emptyset_{TP} = \frac{\emptyset_{PF}\, \rho_M}{1-\emptyset_{PF}+\emptyset_{PF}\, \rho_M} \tag{5}$$

where $\emptyset_{PF}$ is the volume fraction of pore former and ρ_M the packing fraction or relative density of the matrix. Equation (5) assumes that large pores in a fine-grained matrix do not contribute to shrinkage. Actually they shrink according the overall shrinkage of the body that is caused mostly by the matrix.[20] As pores left by graphite are larger than grain size, the porosity related to the burnout of graphite particles is constant during the matrix densification. The idea in the present study was to substitute the packing fraction of the compact body ρ_M with the packing fraction or relative density of suspension. Since Eq. (5) is verified after burnout the relative density of suspension is determined as $\emptyset^*/A$ where A is the maximum solid concentration and $\emptyset^*$ is the volume fraction of powders without pore former fraction. Porosity values are reported in Table V and the difference between measured and calculated porosity results negligible. In order to confirm this finding about the porosity values, percolation expression is invoked:[17]

$$p_C = (\text{packing fraction, } \emptyset/A)\ (\text{percolating phase, } \emptyset_{pf}) \tag{6}$$

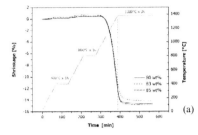 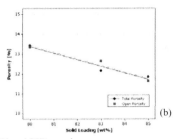

Figure 7. Dilatometric curves (a) and Porosity (b) for 80, 83 and 85D samples.

Equation (6) supplies the critical percolation value corresponding to the porosity value for which percolation is achieved. It is calculated from rheological data and suspension composition through slurry packing fraction and pore former fraction, respectively. In Table V the critical percolation values are nearly 13.9, 13.6 and 14.0% for 80, 83 and 85D samples. These values are slightly higher than measured open porosity. When the open porosity coincides to the critical percolation value one can assess the interconnection of pores and a percolating pore network can be expected. The ratio between open porosity and percolation value estimates the fraction of interconnected porosity and this corresponds to 83.1, 92.9 and 96.0% for 85, 83 and 80D samples, respectively. This means that for 80D sample most of porosity is interconnected even for low porosity value such as around 13%. These limited percolation values are lower than 16% usually encountered for mono-disperse systems when no lattice is present as for powder systems undergoing sintering.[17] Since the percolation value depends on the particle aspect ratio, the low p_c values can be ascribed to the anisotropic shape of pore former. In conclusion, for the system analysed it is expected to obtain a percolating pore network at low porosity value and this is possible through a high packing fraction as well the anisotropic shape of graphite particles.

Table V. High concentrated suspensions: volume fraction of powders without pore former ($\emptyset^*$), maximum solid concentration (A), packing fraction ($\emptyset^*$/A), total porosity calculated by Slamovich-Lange expression ($\emptyset_{TP}$), open porosity ($\emptyset_{OP}$), fraction of interconnected porosity ($\emptyset_{OP}/p_C$) and linear shrinkage.

Suspension [wt%]	$\emptyset^*$ (without pore former)	A	$\emptyset^*$/A	$\emptyset_{TP}$ [%]	$\emptyset_{OP}$ [%]	p_C [%]	$\emptyset_{OP}/p_C$ [%]	Linear shrinkage [%]
80	.3829	.4395	.8712	13.44	13.34 ± .04	13.9	96.0	-14.69
83	.3920	.4867	.8054	12.15	12.63 ± .12	13.6	92.9	-14.14
85	.3979	.4962	.8019	11.84	11.64 ± .01	14.0	83.1	-14.89

CONCLUSIONS

The stability of colloidal suspensions is obtained by good dispersion of slurry systems through a precise control of dispersant amount. The enhanced stability of colloidal slurries improves the control and the reliability of wet colloidal process and minimizes the influence of external conditions such as humidity and temperature. Moreover, the good dispersion of suspensions allows to obtain homogeneous microstructure of the fired bodies as displayed for samples derived from 80 and 83 wt% suspensions. On the other hand high packing efficiency of highly concentrated slurries decreases the critical percolation value and percolating pore network can be obtained at low values of porosity such as 13% for sample derived from 80 wt% slurry. This low porosity value implies the realization of anodes with potential improvement on mechanical properties as well electrical conductivity even if further investigations are required to assess transport properties.

REFERENCES

[1] A. Torabi, T. H. Etsell, P. Sarkar, Dip coating fabrication process for micro-tubular SOFCs, *Solid State Ionics*, 192 (2011), 372.
[2] N.M. Sammes, Y. Du, R. Bove, Design and fabrication of a 100W anode supported micro-tubular SOFC stack, *Journal of Power Sources*, 145 (2005), 428–434.
[3] K. Kendall, Progress in Microtubular Solid Oxide Fuel Cells, *Int. J. Appl. Ceram. Technol.*, 7 [1], (2010), 1.
[4] S.P. Jiang, J.G. Love, L. Apateanu, Effect of contact between electrode and current collector on the performance of solid oxide fuel cells, *Solid State Ionics*, 160 (2003,) 15.
[5] R. De la Torre García, V.M. Sglavo, Fabrication of Innovative Compliant Current Collector-Supported Microtubular Solid Oxide Fuel Cells, *Int. J. Appl. Ceram. Technol.*, (2011), 1.
[6] J.P.P. Huijsmans, Ceramics in solid oxide fuel cells, *Current Opinion in Solid State and Materials Science*, 5, (2001), 317.
[7] W.Z. Zhu, S.C. Deevi, A review on the status of anode materials for solid oxide fuel cells, *Materials Science and Engineering*, A362, (2003), 228.
[8] H.A. Barnes, J.F. Hutton and K. Walters, An Introduction to Rheology, *Rheology Series 3, Elvesier*, 1989.
[9] D. Liu, W.J. Tseng, Rheology of injection-molded zirconia-wax mixtures, *Journal of Materials Science*, 35, (2000), 1009.
[10] P. Mikulášek, R.J. Wakeman, J.Q. Marchant, The influence of pH and temperature on the rheology and stability of aqueous titanium dioxide dispersions, *Chemical Engineering Journal*, 67, (1997), 97.
[11] L. Guo, Y. Zhang, N. Uchida, K. Uematsu, Influence of Temperature on Stability of Aqueous Alumina Slurry Containing Polyelectrolyte Dispersant, *J. Eur. Ceram. Soc.*, 17, (1991), 345.
[12] X. Wang, L. Guo, Effect of temperature on the stability of aqueous ZrO2 suspensions, *Colloids and Surfaces A: Physicochem. Eng. Aspects*, 304, (2007), 1.
[13] D. Liu, Effect of Dispersants on the Rheological Behavior of Zirconia–Wax Suspensions, *J. Am. Ceram. Soc.*, 82, [5], (1999), 1162.
[14] D. Liu, W.J. Tseng, Influence of Powder Agglomerates on the Structure and Rheological Behavior of Injection-Molded Zirconia–Wax Suspensions, *J. Am. Ceram. Soc.*, 82, [10], (1999), 2647.
[15] J. Mewis, N.J. Wagner, Thixotropy, *Adv. Colloidal Interf. Science*, 147–148, (2009), 214.
[16] H.A. Barnes, Thixotropy_a review, *J. Non-Newtonian Fluid Mech.*, 70, (1997), 1.
[17] M.N. Rahaman, Ceramic and Sintering Processing, *Marcel Dekker, 2nd edition*, 2003.
[18] D. Liu, W.J. Tseng, Yield behavior of zirconia-wax suspensions, *Materials Science and Engineering*, A254, (1998), 136.
[19] C.W. Macosko, Rheology Principle, Measurements and Application, *Wiley-VCH*, 1994.
[20] E. Gregorová, W. Pabst, Process control and optimized preparation of porous alumina ceramics by starch consolidation casting, *Journal of the European Ceramic Society*, 31, (2011), 2073.
[21] E.B. Slamovich, F.F. Lange, Densification of Large Pores: I, Experiments, *J. Am. Ceram. Soc.*, 75, [9], (1992), 2498.
[22] J.S. Reed, Principle of Ceramic Processing, *2nd edition, John Wiley & Sons, Inc.*, 1995.
[23] W.J. Tseng, D. Liu, C. Hsu, Influence of stearic acid on suspension structure and green microstructure of injection-molded zirconia ceramics, *Ceram. International*, 25, (1999), 191.
[24] A.J. Burggraaf, Fundamentals of Inorganic Membrane Science and Technology, *Elvesier*, 1996.
[25] R.J. Larson, The Structure and Rheology of Complex Fluids, *Oxford University Press, Inc.*,1999.

[26] I.M. Krieger and T.J. Dougherty, A mechanism for the Non-Newtonian Flow in Suspensions of Rigid Spheres, *Trans. Soc. Rheol.*, III, (1959), 137.

MIXED CONDUCTING PRASEODYMIUM CERIUM GADOLINIUM OXIDE (PCGO) NANO-COMPOSITE CATHODE FOR ITSOFC APPLICATIONS

Rajalekshmi Chockalingam[a], Ashok Kumar Ganguli[b] and Suddhasatwa Basu[a*]
Department of Chemical Engineering[a], Indian Institute of Technology, Delhi,
New Delhi-110016, India,
Department of Chemistry[b], Indian Institute of Technology, Delhi, New Delhi-110 016, India

ABSTRACT

A mixed ionic and electronic conducting (MIEC) cathode material, $Pr_{0.20}Ce_{0.75}Gd_{0.05}O_{2-\delta}$ (PCGO) was synthesized through wet chemical co-precipitation method. Phase analysis of the prepared samples was performed using X-ray diffraction and microstructures of the sintered samples determined by Scanning Electron Microscopy (SEM). Two point and four point DC conductivity measurements were performed to determine the ionic and electronic conductivities of the composite samples as a function of temperature and partial pressures of oxygen in the temperature range 300-800 °C. The XRD results show the formation of a single fluorite phase which appears to be stable up to 900 °C. A pO_2 dependent ionic conductivity was observed at high pO_2 in the PCGO composite at low temperature regions, due to the oxidation of Pr^{3+} to Pr^{4+}. The results of impedance measurements indicate that the area specific resistance (ASR) values increases with increasing Pr content. The ASR value of PCGO is found to be significantly lower than that of conventional $La_{0.6}Sr_{0.4}Co_{0.2}Fe_{0.8}O_{3-\delta}$ (LSCF). The ASR value of LSCF is found to be 6.3687 Ω-cm^2 at 550 °C whereas the $Ce_{0.75}Pr_{0.20}Gd_{0.05}O_{2-\delta}$ sample shows a value 0.299 Ω-cm^2 at the same temperature. Praseodymium cerium gadolinium nano-composite cathode, GDC electrolyte and NiO-GDC anode SOFC produced maximum power density of 380 mWcm^{-2} which indicates a promising material for the application of intermediate temperature solid oxide fuel cells.

INTRODUCTION

Solid oxide fuel cell (SOFC) is an electrochemical device that directly and efficiently converts chemical energy into electrical energy with fuel flexibility [1]. Although the high operating temperature of SOFC offers several advantages over polymer electrolyte based fuel cells (PEM), it suffers high operating cost and material compatibility issues [2]. As a consequence, significant effort has been devoted to reduce the operating temperature of SOFC. The major challenge in reducing the operating temperature of SOFC is the poor activity of traditional cathode materials for electrochemical reduction of oxygen at the temperature range of 500-700 °C due to high activation energy and slow reaction kinetics for the oxygen reduction reactions [3]. Therefore, the development of cathodes with high performance and stability becomes increasingly critical for reducing the operating temperature of SOFC. The basic function of the cathode in an SOFC is to reduce the oxide atoms to oxygen ions via the overall half cell reaction [1]

$$\frac{1}{2}O_2 + 2\,e^- \rightarrow O^{2-} \qquad (1)$$

An ideal cathode should have high electronic and ionic conductivity, good chemical and thermal compatibility with electrolyte and high catalytic activity for the oxygen reduction reaction [3]. Platinum was the first electrode used to reduce the oxygen on yttria stabilized zirconia more than 100 years ago [4], later a low cost alternative to platinum, $La_{1-x}Sr_xMnO_{3\pm\delta}$ (LSM), a pure electronic conducting perovskite material was introduced [5]. Although, LSM exhibits good performance at high temperature, its performance at intermediate temperature is poor due to the low ionic conductivity. Shao and Haile [1] proposed a new intermediate temperature cathode material $Ba_{0.5}Sr_{6.5}Co_{0.8}Fe_{0.2}O_{3-\delta}$ (BSCF) operated with humidified hydrogen as the fuel and air as the cathode gas, demonstrated a power density of 1,010 mWcm^{-2} at 600 °C and 402 mWcm^{-2} at 500 °C. Lee and Manthiram [6] investigated the electrochemical performance of double perovskite cathode material, $Ln_{1-x}Sr_xCoO_{3-\delta}$ (Ln= Pr, Nd, Sm and Gd) capable of operation at intermediate temperature. They reported that $NdBaCo_{1.25}Cu_{0.75}O_{5+\delta}$ and $GdBaCo_{1.0}Cu_{1.0}O_{5+\delta}$ samples showed improved cathode performances compared to the Cu-free ($x = 0$) samples, while a further increase in Cu content in $NdBaCo_{2-x}Cu_xO_{5+\delta}$ decreased the performance [7]. In the $LnBaCo_{2-x}Cu_x O_{5+\delta}$ (Ln = Nd and Gd) system, $NdBaCo_{1.25}Cu_{0.75}O_{5+\delta}$ and $GdBaCo_{1.0}Cu_{1.0}O_{5+\delta}$ exhibited lower TEC and high catalytic activity [7]. Takasu et al. [8] first time investigated the electrical properties and catalytic activities of mixed conducting $Ce_xPr_{1-x}O_2$ (PCO) and found that PCO could be an active catalyst material for oxygen reduction. Nauer et al. [9] further studied PCO as a cathode material and reported that the electrical conductivity of PCO increases with increase in Pr content. Stefanik and Tuller [10] extensively studied the non-stoichiometry and defect chemistry of PCO. In the present investigation, we have studied the structural and electrical properties of $Pr_{0.20}Ce_{0.75}Gd_{0.05}O_{2-\delta}$ and its use as a cathode material for intermediate temperature solid oxide fuel cell (SOFC) applications.

EXPERIMENTAL PROCEDURE

$Pr_{0.20}Ce_{0.75}Gd_{0.05}O_{2-\delta}$ (PCGO) powder was synthesized using a wet-chemical co-precipitation route. The various steps involved in the preparation of powder is given in Figure 1. Appropriate quantities of cerium nitrate hexa hydrate $Ce(NO_3)_3 \cdot 6H_2O$, praseodymium nitrate hexa hydrate $Pr(NO_3)_3 \cdot 6H_2O$ and gadolinium nitrate hexa hydrate $Gd(NO_3)_3 \cdot 6H_2O$ were dissolved in distilled water to form a 0.3 mol solution. The solution was transformed in to a beaker and heat treated at 60 °C for 1 h under vigorous stirring for intimate mixing of precursor solutions prior to the precipitation of the insoluble phase. Ammonium hydroxide (NH_4OH), 0.1 mol L^{-1} solution was slowly added drop wise and continued the stirring for 5 h to complete the co precipitation reaction. The final pH of the dispersion was measured as 7.5. The mixture was kept at this temperature for 6 h while stirring to form a pale orange precipitate at 60 °C. The resulting dispersion was cooled to room temperature and kept for aging over night. The formed precipitate was washed several times with deionized water and subsequently dried at 80 °C. The resultant nano powder was calcined at 900 °C for 5 h.

Phase analysis of the prepared powders was characterized by X-ray diffraction (X'PERT PRO Panalytical, Netherlands). A Cu-Kα radiation with Ni-filter was used as the radiation source and the generator was set to 40 kV and 30 mA in the 2θ range of 10 to 80° with a step size of 0.02° and a step time of 1 sec. The microstructures of the powder and samples were evaluated using HR-TEM (TECNAI G^2 20, FEI), as well as SEM (EVO 50 Zeiss).

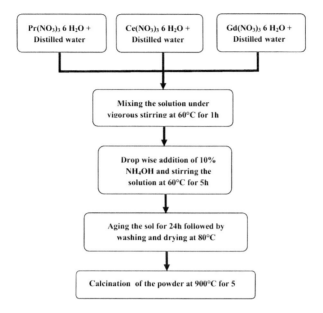

Figure.1 Preparation of $Pr_{0.20}Ce_{0.75}Gd_{0.05}O_{2-\delta}$ (PCGO) nano composite powder using co-precipitation method

A rectangular bar specimen with dimensions of 18 mm length, 5 mm width and 3mm height were prepared for four point DC conductivity measurements. For this purpose appropriate quantity of PCGO powder was loaded into rectangular die and was pressed to 8000 psi using uni axial die press. The resulting pellets were loaded into poly ethylene bags, evacuated using a vacuum pump and heat sealed. The sealed samples were then pressed to 40000 psi pressure using cold iso static press. The samples were then sintered at 1550 °C at a heating rate of 2 °C per minute and 5 h hold. A layer of silver conductive ink thinned by ethyl acetate (Alfa Aesar #42957) was then applied to the bar to form two contacts for two probe technique and four contacts for four probe technique to measure conductivity. Two end contacts were extended to cover the end of the bar, while the two center contacts were simply painted around the exterior. The samples were then wrapped with Ag wire leads and another layer of silver conductive ink was applied to ensure good contact. Each sample was wired for conductivity measurements. The resistance of the PCGO sample was measured using a simple electrochemical cell comprising the PCGO sample with a pair of silver electrodes reversible to the charge carrying species placed on either side. A steady state potential difference was applied between the electrodes and the current response was measured. The conductivity of the material was calculated using the equation given below.

$$\sigma = \left[\frac{I}{V} \frac{L}{A} \right] \qquad (2)$$

Where I is the sample current, V is the voltage drop between the electrodes, A is the cross sectional area and L is the length. In the four point technique the potential difference is measured between two probes using a high impedance voltmeter by applying a current through the sample.

AC impedance measurements were carried out using an AUTOLAB PGSTAT30 FRA. Sintered cylindrical specimens were used for impedance measurements. The specimens were coated on both faces with silver paste and cured at 600 °C with a heating rate of 10 °C/min and 40 min hold. The samples were heated in a muffle furnace and measurements were carried out from 400 to 700 °C. Data were collected with the help of software (FRA). Lead effects can occur due to a measured response from the silver wires used as leads. Correction files were used to eliminate any resistance due to the leads. All data were collected while sweeping from high to low frequency in order to avoid polarization in the sample. Nyquist plots were then generated and the conductivity in S/cm was determined for each composition. The electrolyte supported single cells were fabricated by tape casting the NiO-GDC anode slurry on one side of the electrolyte disk (GDC) and the bilayer was sintered at 1350 °C for 5 hrs at a heating rate of 2 °C per minute and finally the cathode layer (50-50 wt % PCGO-GDC) was brush painted on the rear side of the electrolyte and the single cell was sintered at 1100 °C for 2 h at a heating rate of 2 °C per minute [11,12]. Air with a flow rate of 80 mL/min was supplied to the cathode chamber and humidified H_2 (2% H_2O) at a flow rate of 50 mL/min was the fuel fed to the anode chamber [13].

RESULTS AND DISCUSSION

Figure 2 shows the bright field high resolution transmission electron microscope (HRTEM) image of $Pr_{0.20}Ce_{0.75}Gd_{0.05}O_{2-\delta}$ (PCGO) unagglomerated nano particles. Most of the particles were found to be ~5 nm in diameter and spherical in shape. The X-ray powder diffraction pattern of PCGO powder sample annealed from room temperature to 900 °C is shown in Figure 3. Characteristic diffraction peaks corresponding to typical fluorite like cubic structure was formed after 2 hrs of heat treatment of the powders at 600 °C. The powder completely crystallized at two hours of heat treatment at 900 °C as indicated by the XRD pattern (Figure. 3d). It is interesting to note that peaks corresponding to any oxide of Pr or Gd were not observed. The diffraction patterns are in good agreement with those of pure ceria as indicated by the crystallographic planes in the x-ray pattern (Figure 3d). This result indicates that Pr and Gd had been completely incorporated into the ceria lattice to form a fluorite like $Pr_{0.20}Ce_{0.75}Gd_{0.05}O_{2-\delta}$ (PCGO) phase. The lattice constant, **a**, of PCGO was found to be 5.3854 Å at 80 °C and 5.4235 Å at 900 °C. The mean crystallite size of all samples were calculated by the Scherrer equation based on XRD line widths and are listed in Table 1. The crystallite size of the PCGO samples were calculated as 8.39 and 37.39 nm at 80 °C and 900 °C respectively, which points to the fact that increasing calcination temperature increases the particle size.

Table-I

Temperature (°C)	Lattice Parameter (a) (Å)	Crystallite Size (D) (nm)
80	5.38541	8.385
300	5.41566	8.8997
600	5.42056	12.1726
900	5.42348	37.391

Figure 4 shows the SEM micrograph of fracture surface of PCGO sample sintered at 1550 °C for 5 h. The microstructure clearly shows the formation of grains and no visible porosity is observed.

However, the relative percentage density of the samples calculated by Archimedes principle indicates a value of 92.5 %. There were also reports in the literature showing improved densification behavior of similar composition while the sintering temperature and time were increased. Schmale et al., sintered $Ce_{0.80}Gd_{0.2-x}Pr_xO_{2-\delta}$ composition at 1600 °C for 10 h to obtain full density [14]. At the same time Kharton et al., sintered Pr doped $Ce(Gd)O_{2-\delta}$ at 1600 °C for 6 h and reported a density of 95% [15]. Maffei et al., sintered the composition $Ce_{0.8}Gd_{0.19}Pr_{0.01}O_{1.905}$ at 1600 °C for 12 h and achieved a density of 97% [16]. From their studies, it appears that the samples need to be sintered 1600 °C to obtain full densification.

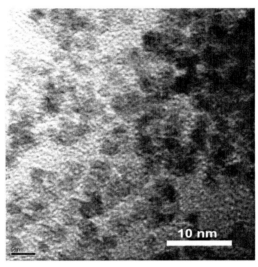

Figure 2. HRTEM image of $Pr_{0.20}Ce_{0.75}Gd_{0.05}O_{2-\delta}$ (PCGO) nano particles syntheszed through co-precipitation method.

The electrical conductivity of PCGO sample was measured isothermally as a function of partial pressure of oxygen (pO_2) at a series of temperatures ranging from 400 to 800 °C. The results are shown in Figure 5. The conductivity of PCGO sample measured as a function of pO_2 shows similar trend irrespective of temperature. The conductivity increases up to 10^{-6} atm of pO_2 and thereafter it shows insensitivity to changes in pO_2. At higher partial pressure of oxygen,

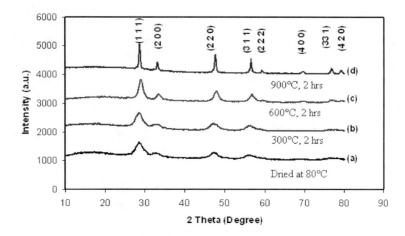

Figure 3. XRD pattern of the as-prepared $Pr_{0.20}Ce_{0.75}Gd_{0.05}O_{2-\delta}$ (PCGO) powders (a) PCGO dried at 80°C (b) PCGO calcined at 300°C for 2 h (c) PCGO calcined at 600°C for 2 h (d) PCGO calcined at 900°C for 2 h respectively.

Figure 4 SEM microstructure showing the fracture surface of $Pr_{0.20}Ce_{0.75}Gd_{0.05}O_{2-\delta}$ sample sintered at 1550 °C for 5 h.

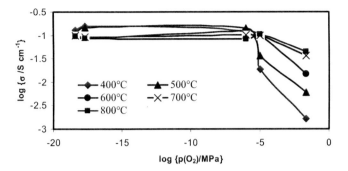

Figure 5. Two point DC conductivity results of $Pr_{0.20}Ce_{0.75}Gd_{0.05}O_{2-\delta}$ samples as a function of partial pressure of oxygen (pO_2) at a series of temperatures ranging from 400 to 800 °C

conductivity increases with increase in temperature. At the same time, the effect of temperature is not significant at a pO_2 of 10^{-5} atm and lower values up to 10^{-18} atm. A maximum conductivity of 0.3 S cm^{-1} was exhibited by the sample measured at 500 °C at 10^{-18} atm of partial pressure of oxygen. In contrast to two point DC conductivity measurements, the conductivity of PCGO sample increases with increase in temperature at all partial pressures of oxygen in the case of four point DC conductivity measurement. The sample measured at 400 and 600 °C shows a -1/4 dependence on pO_2 in the region 10^{-5} to 10^{-18} atm of pO_2. The results indicate that the variation in two point DC conductivity values compared to four point DC at higher pO_2 may be due to the error at higher pO_2 because this technique does not take into account any changes in the composition of the electrode during the experiment and also it assumes that the electrode resistance is negligible compared to the electrolyte resistance. Four point DC conductivity eliminates the contribution due to electrode- electrolyte interface. It is known that CeO_2 is partially reduced from Ce^{4+} to Ce^{3+} at lower pO_2. Due to this reduction, CeO_2 generates electronic charge carriers according to the reaction:

$$O_O = V_O^{\bullet\bullet} + 2\,e' + \frac{1}{2}\,O_2$$

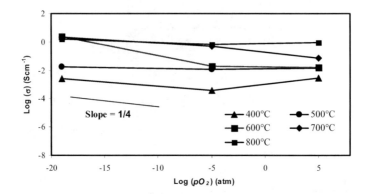

Figure 6. Four point DC conductivity results of $Pr_{0.20}Ce_{0.75}Gd_{0.05}O_{2-\delta}$ samples as a function of pO_2 at a series of temperatures ranging from 400 to 800 °C.

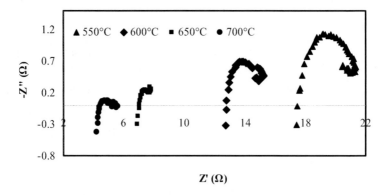

Figure 7. Complex AC-impedance spectra of $Pr_{0.20}Ce_{0.75}Gd_{0.05}O_{2-\delta}$ samples measured at various temperatures.

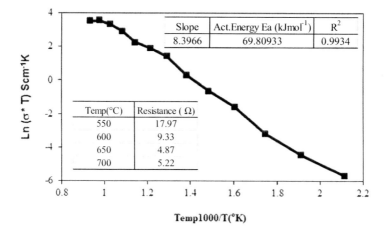

Slope	Act.Energy Ea (kJmol^{-1})	R^2
8.3966	69.80933	0.9934

Temp(°C)	Resistance (Ω)
550	17.97
600	9.33
650	4.87
700	5.22

Temp1000/T(°K)

Figure 8. Total conductivity of PCGO as a function of reciprocal temperature

Figure 7 shows the complex impedance spectra of PCGO sample measured at various temperatures. A single but slightly depressed semi circle appeared on the impedance plots of cathodes at open circuit conditions. Cathode polarization resistance estimated by impedance measurement decreases with increase in temperature. The vertical part below the real axis in all the plots is due to the inductive elements in the experimental set up. The single semi circle at all temperatures clearly implies that electronic conductivity is much larger than the ionic one. The absence of third semi circle indicates that the electrode resistance is much larger than the electronic resistance [9]. While fitting the curve to the equivalent circuit, Warburg diffusion impedance was included in the low frequency region. The results of impedance measurements indicate that the area specific resistance (ASR) values increases with increasing Pr content. The ASR value of PCGO is found to be significantly lower than that of the value of LSCF quoted on the existing literature [17]. Our experimental results demonstrated an ASR value of 6.3687 Ω-cm^2 at 550 °C for LSCF samples and 0.299 Ω-cm^2 for $Ce_{0.75}Pr_{0.20}Gd_{0.05}O_{2-\delta}$ at the same temperature. Figure 8 shows total conductivity of PCGO as a function of reciprocal temperature. As temperature increases, conductivity also increases. The total conductivity of $Ce_{0.75}Pr_{0.20}Gd_{0.05}O_{2-\delta}$ increased from 0.0053 to 0.04983 S cm^{-1} when the temperature is linearly increased from 500 to 650 °C. The activation energy of PCGO estimated from the plot is 69.81 kJ mol^{-1} at a temperature range of 200-800 °C and 23.39 kJ mol^{-1} at a temperature range of 500-650 °C. Whereas the total conductivity of LSCF sample varies from 0.0199 to 0.0419 S cm^{-1} at temperature ranging from 500 to 650 °C. The activation energy is estimated as 36.912 kJ mol^{-1} which is lower than that of the value, 142 kJ mol^{-1} reported in the literature by Hwang et al. at the same temperature range [18].

In order to evaluate the electrochemical performance single cells were fabricated using $Ce_{0.80}Gd_{0.20}O_2$ as electrolyte, 42 vol % NiO and 58 vol % GDC as anode and 50:50 wt. % $Pr_{0.20}Ce_{0.75}Gd_{0.05}O_{2-\delta}$-GDC cathode. Polystyrene was used as a pore former in both the electrode mixtures. The electrolyte was prepared by die pressing. The 4 mg of GDC electrolyte powder was filled in a 13 mm die and uniaxially pressed at 7000 PSI followed by cold iso static pressing at 40000 PSI. The cathode ink was prepared by mixing of $Ce_{0.75}Pr_{0.20}Gd_{0.05}O_{2-\delta}$ (PCGO) and GDC powder in a 50:50 wt %, mixed thoroughly together with a 6 wt. % of ethyl cellulose binder and the resulting mixture was ball milled with a terpineol based solvent for 12 h. Anode was tape casted on the electrolyte disks and cathode was brush painted on the other side of the electrolyte. The single cell was sintered at 1450 °C for 10 h at a heating rate of 2 °C per minute. The thickness of the anode, electrolyte and cathode were approximately 1, 1 and 0.5 mm respectively after sintering. Anode surface of the sintered cell was covered with nickel gauze and cathode surface was covered with silver gauze as current collectors, which were then coated with silver conductive ink as sealant. Silver wires were used as the connection leads and heat treated at 200 °C for 1 h for a better electrical contact. The single button cell was mounted in between the air and fuel chambers of an indigenously developed stainless steel casing and the whole cell set up was connected to the fuel cell test station. The cells were tested between 450 to 650 °C with humidified hydrogen gas as the fuel and zero air as oxidant. Flow rates of gases were controlled between 40 and 100 mL min^{-1} at one atmosphere pressure. The performance analysis of the fuel cells was carried out using potentistat/galvanostat (Autolab, PGSTAT 30).

The measurements were performed from 450 °C to 650 °C with humidified hydrogen (2% H_2O) as fuel and air as oxidant. Figure 9 compares the I-V curves and power density of single cell at three temperatures 450, 550 and 650 °C respectively. The power density increased with increasing temperature while OCV remains the same and a maximum power density of 380 mWcm^{-2} is obtained at 650 °C. The performance of the cathode can be explained by the combined ionic and electronic contributions of MIEC. PCGO is responsible for electronic conductivity while GDC contribute ionic conductivity. GDC cathode layer has high activity for oxygen reduction and oxygen ion migration, while oxygen ion migration is the rate limiting step for cathode performance.

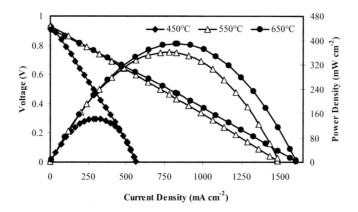

Figure 9 performance analysis of a single cell fabricated with NiO-GDC anode, GDC electrolyte and 50-50 wt. % PCGO-GDC cathode at a temperature range of 450 -650 °C.

CONCLUSIONS

Mixed electronic and ionic conducting $Pr_{0.20}Ce_{0.75}Gd_{0.05}O_{2-\delta}$ has been evaluated as a cathode material for intermediate temperature solid oxide fuel cell applications. Four point DC conductivity results showed ¼ dependence of pO_2, indicating electronic conductivity at 400 and 600 °C. The total conductivity increased with increasing temperature measured by AC impedance spectroscopy. Maximum conductivity of 0.0053 Scm^{-1} was observed at 500 °C. The activation energy estimated from the plot is 69.81 kJ mol^{-1} at temperature range of 500-800 °C. PCGO exhibited a maximum power density of 380 mW cm^{-2} at 650 °C.

ACKNOWLEDGEMENTS

This work was supported by CSIR, India (Project No: 22 (0524)/10/EMR-II). Authors gratefully acknowledge Indian Institute of Technology, Delhi for financial assistance for graduate studies, National Physical Laboratory, New Delhi, India for technical assistance and Department of Science and Technology, New Delhi, India for Travel Grant (No.SR/ITS/4271/2011-2012).

REFERENCES

1. Z. Shao and S. M. Haile, "A High Performance Cathode for the Next Generation Solid-Oxide Fuel Cells," *Nature* 431, 170-173 (2004).
2. B. C. H. Steele and Angelika Heinzel,"Materials for fuel-cell technologies," *Nature* Volume 414, 345-352 (2001).
3. S. B. Adler. "Factors Governing Oxygen Reduction in Solid Oxide Fuel cell Cathodes," *Chem. Rev.*, 104, pp: 4791-4843 (2004).
4. H. H. Mobius, "On the History of Solid Electrolytes", *J. Solid State Elecrochemistry*, 1, pp 2-16 (1997).
5. Solid Oxide Fuel Cells Materials Properties and Performance, edited by Jeffrey W. Fergus, Rob Hui, Xianguo Li, David P. Wilkinson and Jiujun Zhang, CRC Press, Taylor&Francs Group,LLC, (2009).
6. K. T. Lee and A. Manthiram, "Comparison of $Ln_{0.6}Sr_{0.4}CoO_{3-\delta}$ (Ln=La, Pr, Nd, Sm, and Gd) as Cathode Materials for Intermediate Temperature Solid Oxide Fuel Cells," *J. Electrochem. Soc.*, Volume 153, Issue 4, pp. A794-A798 (2006).
7. Y. N. Kim and A. Manthiram, "Layered $LnBaCo_{2-x}Cu_xO_{5+\delta}$ ($0 \leqslant x \leqslant 1.0$) Perovskite Cathodes for Intermediate-Temperature Solid Oxide Fuel Cells,"*J. Electrochem. Soc.,* Volume 158, Issue 3, pp. B276-B282 (2011).
8. Y. Takasu, T. Sugino and Y. Matsuda, "Electrical conductivity of praseodymia doped ceria," *J. Appl. Electrochem.,* 14, 79-81 (1984).
9. M. Nauer, C. Ftikos and B. C. H. Steele, "An Evaluation of Ce-Pr Oxides and Ce-Pr-Nb Oxides Mixed Conductors for cathodes of Solid Oxide Fuel cells: Structure, Thermal Expansion and Electrical Conductivity", *J. Euro Ceram Soc.* 14,493-499 (1994).
10. T. S. Stefanik and H. L. Tuller, "Nonstoichiometry and Defect Chemistry in Praseodymium-Cerium Oxide", *J. Electroceramics,* 13, 799-803, (2004).
11. C. Fu, S. H. Chan, Q. Liu, X. Ge and G. Pasciak, "Fabrication and evaluation of Ni-GDC composite anode prepared by aqueous-based tape casting method for low-temperature solid oxide fuel cell", *Int. J. Hydrogen Energy.*, 35, 301-307 (2010).
12. C. Zhu, X. Liu, C. Yi, L. Pei, D. Wang, D. Yan, K. Yao, T. Lu and W. Su, "High-performance $PrBaCo_2O_{5+\delta}$-$Ce_{0.8}Sm_{0.2}O_{1.9}$ composite cathodes for intermediate temperature solid oxide fuel cell", *J. Power Sources.*, 195, 3504-3507 (2010).

13. W. Sun, Z. Shi, S. Fang, L. Yan, Z. Zhu and W. Liu, "A high performance $BaZr_{0.1}Ce_{0.7}Y_{0.2}O_{3-\delta}$-based solid oxide fuel cell with a cobalt-free $Ba_{0.5}Sr_{0.5}FeO_{3-\delta}$-$Ce_{0.8}Sm_{0.2}O_{2-\delta}$ composite cathode", *Int. J. Hydrogen Energy.*, 35, 301-307 (2010).

14. K. Schmale, M. Grunebaum, M. Janssen, S. Baumann, F. Schulze-Kuppers and Hans-Dieter Wiemhofer, "Electronic conductivity of $Ce_{0.80}Gd_{0.2-x}Pr_xO_{2-\delta}$ and influence of added CoO", *Phys. Status Solidi B* 248, No. 2, 314-322 (2011).

15. V. V. Kharton, A. P.Viskup, F. M. Figueiredo, E. N. Naumovich, A. A. Yaremchenko and F.M.B. Marques," Electron-hole conduction in Pr-doped $Ce(Gd)O_{2-\delta}$ by faradaic efficiency and e m f measurements", *Electrochimica Acta* 46 2879-2889 (2001).

16. N. Maffei, and A.K. Kuriakose, "Solid oxide fuel cells of ceria doped with gadolinium and praseodymium", *Solid State Ionics*, 107 67-71 (1998).

17. S. P. Jiang, "A comparison of O_2 reduction reactions on porous (La,Sr)MnO$_3$ and (La,Sr)(Co,Fe)O$_3$ electrodes," *Solid State Ionics*, 146, 1-22 (2002).

18. H. J. Hwang, J. W. Moon, S. Lee and E.A. Lee, "Electrochemical performance of LSCF-based composite cathodes for intermediate temperature SOFCs", *J. Power Sources*; 145: 243-248 (2005).

DEVELOPMENT OF GDC-(LiNa)CO$_3$ NANO-COMPOSITE ELECTROLYTES FOR LOW TEMPERATURE SOLID OXIDE FUEL CELLS

Rajalekshmi Chockalingam[a], Ashok Kumar Ganguli[b] and Suddhasatwa Basu*[a]
Department of Chemical Engineering[a], Indian Institute of Technology, New Delhi-110 016, India, Phone: 91-11-2659 1035, Fax: 91-11-2685 1169, Email:sbasu@chemical.iitd.ac.in
Department of Chemistry[b], Indian Institute of Technology, New Delhi-110 016, India

ABSTRACT

Nano composite electrolytes made of gadolinium doped ceria and a bi-phase mixture of lithium carbonate and sodium carbonate salts are investigated with respect to their crystal structure, morphology and electrical conductivity. The addition of mixed salts reduced the sintering temperature of gadolinium doped ceria by forming a eutectic melt at 497 °C and improved the ionic conductivity especially at low temperature. The results derived show formation of a single fluorite phase which appears to be stable up to 800 °C. Electrochemical impedance spectroscopy (EIS) measurements were performed to estimate the total ionic conductivity of gadolinium doped ceria with varying amounts of lithium-sodium carbonate and the results show that the ionic conductivity increased up to 25 wt% carbonate and there after decreased. The typical nature of the impedance spectra of the composite shows the possibility of coexistence of multi ionic transport or existence of space charge effect at the interface of Gd-CeO$_2$/(LiNa)CO$_3$ phase. The composite containing 25 wt% carbonate shows conductivity ranging from 0.1751-0.239 S cm^{-1} within the temperature range of 400-600 °C. Single cells were fabricated with NiO-GDC-25 wt% (LiNa) CO$_3$ as anode || GDC-(LiNa)CO$_3$ electrolyte || LSCF (La$_{0.6}$Sr$_{0.4}$Co$_{0.2}$Fe$_{0.8}$O$_{3-\delta}$)-25 wt% (LiNa)CO$_3$ cathode, tested at 500°C using hydrogen as fuel and air as oxidant delivers a maximum power density of 500 mW/cm^2.

INTRODUCTION

There exists lot of interest in fuel cell community to reduce the operating temperature of solid oxide fuel cells (SOFC) [1, 2]. Recently, Ceres Power, a UK based company developed gadolinium doped ceria based intermediate temperature solid oxide fuel cell capable of operation at 500-600 °C [3]. The use of stainless steel substrate allows one to seal the stacks through conventional welding techniques and have thermal expansion coefficient well matched to their ceramic coatings [3]. Reducing the operating temperature of SOFC also allows greater flexibility in the choice of fuel which in turn reduces the operational cost. There are many reports describing enhanced ionic conductivity, in cationic doped ceria-carbonate composite materials such as gadolinia doped ceria (GDC)-NaCl, GDC-NaOH, GDC-MCO$_3$ (M = Ca, Ba, Sr), Samaria doped ceria (SDC)-M$_2$CO$_3$ (M = Li, Na, K) [4]. The core-shell nanostructured SDC-Na$_2$CO$_3$ composite demonstrated power density of 200-1150 mW/cm^2 at 300-600 °C [5]. The enhanced performance of the fuel cell was explained as an interfacial superionic conduction mechanism in the two phase region which is however questionable due to inadequate proof. The higher defect concentration in the amorphous Na$_2$CO$_3$ phase contribute significantly to Na$^+$-O^{2-} interactions and facilitate the oxygen ion transportation through the interfacial mechanism and Na$_2$CO$_3$ acted as a diffusion barrier layer to reduce the electronic conduction of CeO$_2$ by suppressing the reduction of Ce^{4+} to Ce^{3+} in hydrogen atmosphere above 450 °C [6].

In our previous work we have studied the phase formation, micro structural evolution and ionic conductivity of three compositions: (1) GdCeO$_2$-(LiNa)CO$_3$, (2) Gd-CeO$_2$-Li$_2$CO$_3$ and (3) Gd-CeO$_2$-Na$_2$CO$_3$ [7]. We further investigated the electrical performance of GdCeO$_2$-25 wt.% (LiNa)CO$_3$ single symmetrical cells with NiO-CuO-GDC 25 wt% (LiNa)CO$_3$ anode and lithiated NiO- GDC 25 wt% (LiNa)CO$_3$ cathode fabricated through die-pressing technique [8]. The cell delivered a maximum power density of 92 mW cm^{-2} at 550 °C [8]. The poor performance of the cell was attributed to the porosity of the electrolyte. In the present work GdCeO$_2$-25 wt. % (LiNa)CO$_3$ electrolyte was cold isostatically pressed followed by die-pressing and sintering to full density. Instead of using symmetric cell, NiO-GdCeO$_2$-25 wt. % (LiNa)CO$_3$ anode was tape casted on the sintered electrolyte and La$_{0.6}$Sr$_{0.4}$Co$_{0.2}$Fe$_{0.8}$O$_{3-\delta}$ (LSCF) - GdCeO$_2$-25 wt.% (LiNa)CO$_3$ cathode was brush painted on the other side of the electrolyte. The effects of isostatic pressing on structural and electrical properties were investigated. Attempts also have been made to compare the performance of GDC-25% (LiNa)CO$_3$ with that of SDC-25% (LiNa)CO$_3$ available in literature [9].

EXPERIMENTAL PROCEDURE

In order to prepare GDC-(LiNa)$_2$CO$_3$ nano composite electrolytes, first the carbonate powders were prepared by mixing 53 mol% Li$_2$CO$_3$ and 47 mol% Na$_2$CO$_3$ (Merck Specialities Pvt Ltd.). The carbonate mixture was then mixed with 20 mol% Gd doped CeO$_2$ nano powder prepared by a wet chemical co-precipitation method. The details of the preparation are reported elsewhere [10]. Five compositions of GDC-(LiNa)$_2$CO$_3$ are prepared by varying the carbonate content which are given in Table 1. The mixture of the GDC-carbonates composite powder was then ground thoroughly using a mortar and pestle and heat treated at 680°C in air for 40 minutes. The resultant mixture after heat treatment was again ground into a fine powder with a mortar and pestle and sieved using a test sieve of mesh size of 40 μm. The composite powder was pressed into 13 mm diameter 2 mm thick compacts via dry pressing at 7000 PSI by uni axial die pressing followed by cold isostatic pressing with an applied pressure of 40,000 PSI. The green pellets were sintered at 600 °C for 1 hour.

Table-1 The details of the compositions prepared.

Sample #	Wt % GDC	Wt % (LiNa)CO$_3$
1	90	10
2	85	15
3	80	20
4	75	25
5	70	30
6	65	35

The microstructures of the sintered samples were evaluated using scanning electron microscopy (SEM). Phase analysis was carried out by X-ray diffraction (X'PERT PRO Panalytical, Netherlands). The generator was set to 40 kV and 30 mA utilizing Cu-Kα radiation. Differential thermal analysis (DTA) and thermo gravimetric analysis (TGA) were performed using NETZSCH–TA45 simultaneous DTA-TGA. The data acquisition and processing were carried out using the software Proteus Analysis.

AC impedance measurements were carried out using a frequency response analyzer (AUTOLAB PGSTAT30 FRA). Sintered cylindrical specimens were used for impedance measurements. The specimens were coated on both faces with silver paste and cured at 400 °C with a heating rate of 10 °C/min and held for 40 min. The samples were heated in a muffle furnace and measurements were carried out from 400 to 700 °C. Lead effects can occur due to a measured response from the silver wires used as leads. Correction files were used to account for and eliminate any resistance due to the leads. All data were collected while sweeping from high to low frequency in order to avoid polarization in the sample. Nyquist plots were generated and the conductivity in S/cm was determined for each composition. Single cells were fabricated using GDC-25 wt% (LiNa)CO$_3$ as electrolyte, NiO-GDC-25 wt% (LiNa)CO$_3$ as anode and La$_{0.6}$Sr$_{0.4}$Co$_{0.2}$Fe$_{0.8}$O$_{3-\delta}$ (LSCF)-25 wt% (LiNa)CO$_3$ as cathode. Polystyrene was used as a pore former in both the electrode mixtures. The 13 mm die was filled with electrolyte powder and uniaxially pressed at 7000 PSI followed by cold iso static pressing at 40000 PSI. Anode was tape casted simultaneously on 20 electrolyte samples arranged side by side. Once the anode-electrolyte bi layer was dried, the cathode was brush painted on the other side of the electrolyte. The cathode ink was prepared by mixing $La_{0.6}Sr_{0.4}Co_{0.2}Fe_{0.8}O_{3-\delta}$ (LSCF) and GDC-25%LiNaCO₃ powder in a 7:3 weight ratio, mixed thoroughly together with a 6 wt. % of ethyl cellulose binder and the resulting mixture was ball milled with a terpineol based solvent for 12h and brush painted on the electrolyte surface of the anode/electrolyte bilayer, dried and finally the single cell was sintered at 680 °C for 1 h. Anode surface of the sintered cell was covered with nickel gauze and cathode surface was covered with silver gauze as current collectors, which were then coated with silver conductive ink as sealant. Silver wires were used as the connection leads and heat treated at 200 °C for 1 h for a better electrical contact. The single button cell was mounted in between the air and fuel chambers of an indigenously developed stainless steel casing and the whole cell set up was connected to the fuel cell test station. The cells were tested between 400 to 600 °C with humidified hydrogen gas as the fuel and zero air as oxidant. Flow rates of gases were controlled between 40 and 100 mL min⁻¹ at one atmosphere pressure. The performance analysis of the fuel cells was carried out using Potentistat/Galvanostat (Autolab, PGSTAT 30).

RESULTS AND DISCUSSION

Figure 1 shows the thermogravimetric and differential thermal analysis (TGA and DTA) of the Gd-CeO₂-(LiNa) CO₃ powder. The TGA exhibited three weight loss steps at 75, 180 and 320 °C. The first, second and third weight loss steps are accompanied by three endothermic peaks at 75, 180 and 320 °C. The first endothermic peak at 75 °C is attributed to loss of molecular water adsorbed on the surface of the powder. The second endothermic peak at 180 °C is presumably due to dehydration of more strongly chemisorbed water and the third peak at 320 °C is attributed to the decomposition of cerium based hydrates. Figure 2 shows XRD phase analysis of the sample Gd-CeO₂-(LiNa)CO₃ sintered at 600 °C. The major phase identified is Gd-CeO₂ and minor phase is LiNaCO₃. The presence of carbon and sodium are confirmed by EDX analysis as shown in Figure 4. The EDX results point to the fact that part of the carbonate phase may be present as amorphous state in the case of Gd-CeO₂-(LiNa)CO₃ samples. The SEM images designated in figure 3 (a) to (f) show the microstructures of GdCeO₂ with 10, 15, 20, 25, 30 and 35 wt% (LiNa)CO₃, respectively.

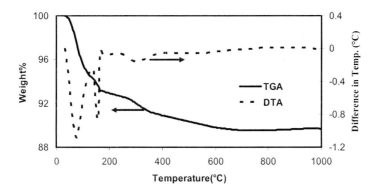

Figure 1. TGA and DTA analysis of synthesized Gd-CeO₂-25% (LiNa) CO₃ powder

All compositions except 35 wt% carbonate show dense microstructure without any visible porosity. A distinct grain formation is observed only in the case of 25 wt% carbonate samples, suggesting that secondary crystallization could have been occurred. In the process of re-crystallization, a few large crystals were nucleated and grew at the expense of fine grains but essentially a strain free matrix was observed [11]. Other compositions show a continuous amorphous like microstructure.

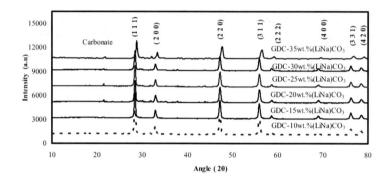

Figure 2. Xray diffraction patterns of GDC-carbonate composites with varying amount of (LiNa)CO₃ sintered at 600 °C for 1h.

The ac-impedance measurements were performed in air by varying the sample temperature from 550 to 700 °C with the help of a split furnace. The data was collected at regular intervals (50 °C) of temperature. Figure 5 shows the typical impedance plots of six samples (GDC-10%(LiNa)CO₃, GDC-15%(LiNa)CO₃ GDC-20%(LiNa)CO₃, GDC-25% (LiNa)CO₃, GDC-30% (LiNa)CO₃ and GDC-35% (LiNa)CO₃) designated as 5(a) to (f) measured at temperatures 550, 600, 650 and 700 °C, respectively. The impedance spectra of GDC with 10 wt. % of carbonate sample (Figure 5a) shows typically a high frequency incomplete depressed semi circle, suggesting that electronic conductivity is

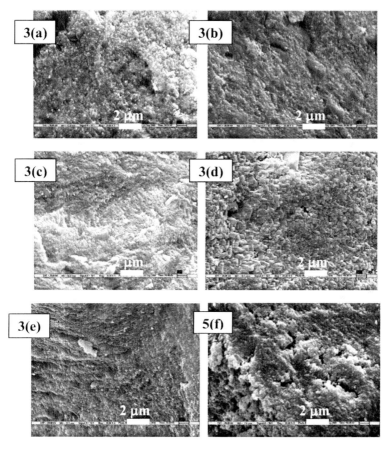

Figure 3. SEM micro graphs of sintered GDC-carbonate composite samples.(a) Gd-CeO₂-10% (LiNa) CO₃ , (b) Gd-CeO₂-15% (LiNa) CO₃ (c) Gd-CeO₂-20% (LiNa) CO₃ (d) Gd-CeO₂-25% (LiNa) CO₃ (e) Gd-CeO₂-30% (LiNa) CO₃ (f) Gd-CeO₂-35% (LiNa) CO₃ sintered at 600 °C for 1h hold.

much larger than the ionic one. The absence of third semicircle indicates that electrode resistance is much larger than the electronic resistance [12]. When the temperature is increased, the resistance decreased. Lapa et al. [13] and Tang et al. [14] reported similar impedance spectra for samaria doped ceria and (Na/Li)CO₃. They argued that the presence of one single high frequency semi circle corresponds to composite effect due to the coexistence of several processes with different relaxation times.

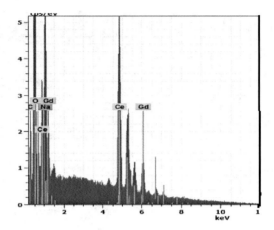

Figure 4. EDX analysis of GDC-25 wt% (LiNa) CO₃ sample showing the presence of Na.

Similar behavior is observed in the case of 15, 20 and 25 wt. % of carbonate samples at 650 and 700 °C. However, the resistance is higher at 550 and 600 °C. The 25 wt% carbonate sample exhibited lowest resistance value of 1.007 Ω at 550 °C. Temperature has no effect on the resistance value as seen in the case of 10 wt% carbonate sample. When the carbonate content increased beyond 25 wt%, the resistance increased with increasing temperature as well as carbonate content. Figure 6. Shows the conductivity of six composite samples of GDC with (LiNa)CO₃ varying from 10-35 weight.% designated as (a) Gd-CeO₂-10% (LiNa) CO₃ , (b) Gd-CeO₂-15% (LiNa) CO₃ (c) Gd-CeO₂-20% (LiNa) CO₃ (d) Gd-CeO₂-25% (LiNa) CO₃ (e) Gd-CeO₂-30% (LiNa) CO₃ (f) Gd-CeO₂-35% (LiNa) CO₃ sintered at 600 °C for 1h hold and it has been observed that the total conductivity of GDC-Carbonate composite increases with increasing temperature.

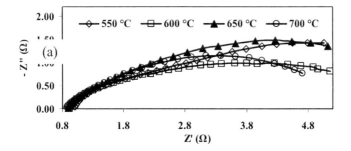

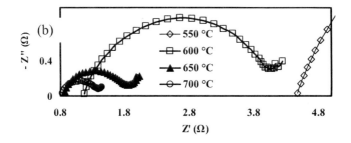

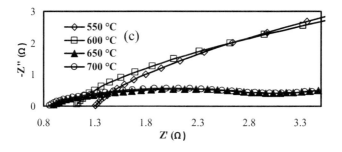

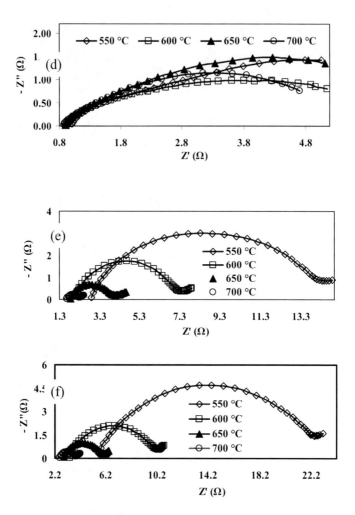

Figure 5. The typical impedance plots of.(a) Gd-CeO$_2$-10% (LiNa) CO$_3$, (b) Gd-CeO$_2$-15% (LiNa) CO$_3$ (c) Gd-CeO$_2$-20% (LiNa) CO$_3$ (d) Gd-CeO$_2$-25% (LiNa) CO$_3$ (e) Gd-CeO$_2$-30% (LiNa) CO$_3$ (f) Gd-CeO$_2$-35% (LiNa) CO$_3$ sintered at 600 °C for 1h hold.

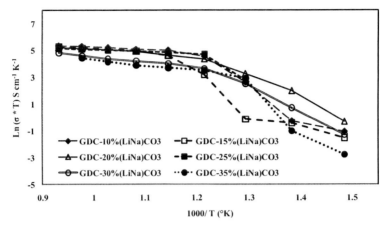

Figure 6. The effect of temperature on the conductivity of.samples (a) Gd-CeO₂-10% (LiNa) CO₃ (b) Gd-CeO₂-15% (LiNa) CO₃ (c) Gd-CeO₂-20% (LiNa) CO₃ (d) Gd-CeO₂-25% (LiNa) CO₃ (e) Gd-CeO₂-30% (LiNa) CO₃ (f) Gd-CeO₂-35% (LiNa) CO₃ sintered at 600 °C for 1h hold.

Table 2 shows the conductivity of all the six compositions at a temperature ranging from 550-700°C. The sample which contains 25 wt% carbonate exhibited highest conductivity of 0.132 Scm⁻¹ at 550 °C, followed by 10 wt% carbonate sample (0.114 Scm⁻¹). The highest conductivity observed in the case of GdCeO₂-25 wt% (LiNa)CO₃ sample could be explained based on the melting transition of eutectic composition of 53 mol% Li₂CO₃ and 47 mol% Na₂CO₃ which in turn created more defects leading to enhanced ionic conductivity. Similar observations have been reported previously by Maier (1995) [15]. An equilibrium concentration of point defects always occurs in any ionic crystal due to the increase of configurational entropy. Then a transition occurs from a low disordered state to a high disordered state. One can observe such transition in the case of AgI. Based on the above fact, one can argue that the GDC-Carbonate two phase regions create a space charge region and a local phase transition occurs at the boundary. Small activation energy in this region creates a liquid like surface conduction leading to higher conductivity [16].

Table: 2 Total conductivity of GDC with varying carbonate content at various temperatures

Sr No	Composition	Temperature (°C)	Conductivity(S/cm²)
1	GDC-10 % LiNa CO₃	550	0.114
		600	0.169
		650	0.176
		700	0.184
2	GDC-15 % LiNa CO₃	550	0.030
		600	0.113
		650	0.153
		700	0.160
3	GDC-20 % LiNa CO₃	550	0.093
		600	0.114
		650	0.148
		700	0.155
4	GDC-25 % LiNa CO₃	550	0.132
		600	0.144
		650	0.149
		700	0.152
5	GDC-30 % LiNa CO₃	550	0.046
		600	0.062
		650	0.071
		700	0.080
6	GDC-35 % LiNa CO₃	550	0.039
		600	0.045
		650	0.052
		700	0.065

An alternative explanation for higher ionic conduction may be caused by the reduction in electronic conductivity of ceria. A similar observation has previously been reported in the case of CeO_2-Al_2O_3 composites [10]. High conductivity could may also be caused by hybrid conduction of proton (H^+) and oxygen ion (O^{--}). Table 3 shows the activation energy of the six compositions of the GDC-(LiNa)CO₃ calculated in the temperature range of 400-800 °C. The activation energy of the composite samples is in good agreement with the conductivity results. The lowest activation energy of 0.42 kJmol⁻¹ is shown by the GDC-25 wt% carbonate sample and highest activation energy of 1.17 kJmol⁻¹ is shown by the GDC-15 wt% carbonate sample.

Table: 3. Activation energy of GDC samples with various amount of (LiNa)CO$_3$ estimated at a temperature range of 400-800 °C

Sample	Temperature (°C)	Slope	E$_a$ (kJ mol^{-1})	R^2
GDC-10% (LiNa)CO$_3$	400-800	12.28	1.02	0.842
GDC-15% (LiNa)CO$_3$	400-800	14.04	1.17	0.899
GDC-20% (LiNa)CO$_3$	400-800	9.171	0.76	0.857
GDC-25% (LiNa)CO$_3$	500-800	5.081	0.42	0.618
GDC-30% (LiNa)CO$_3$	400-800	10.29	0.86	0.881
GDC-35% (LiNa)CO$_3$	400-750	13.75	1.14	0.824

Figure 7 shows the performance analysis of a single cell fabricated with NiO-GDC-25 wt% (LiNa)CO$_3$ anode ‖ GDC-(LiNa)CO$_3$ electrolyte ‖ LSCF (La$_{0.6}$Sr$_{0.4}$Co$_{0.2}$Fe$_{0.8}$O$_{3-\delta}$)-25 wt% (LiNa)CO$_3$ cathode at temperatures ranging from 500-700 °C. Although a maximum open circuit voltage of 0.94V was measured at 600 °C, the current density was 1074 mA cm^{-2} and power density was 267 mW cm^{-2} at this temperature. A detailed description of the open circuit voltage, current density and power density are given in table 4. Here the maximum power densities of 129, 330, and 525 mW cm^{-2} at 500, 550 and 600 °C are comparable with the maximum power densities of 320, 420 and 510 mW cm^{-2} respectively at the same temperature reported by J. Huang et al., with NiO-SDC-(LiNa)CO$_3$ anode ‖ SDC-(LiNa)CO$_3$ electrolyte ‖ Lithiated NiO-SDC- (LiNa)CO$_3$ based fuel cell [9].

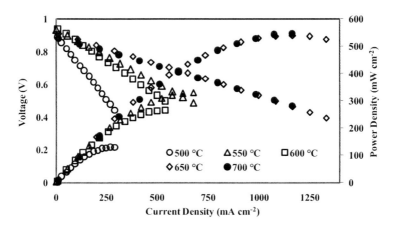

Figure 7. Performance analysis of single cells fabricated with NiO-GDC-25 wt% (LiNa)CO$_3$ anode ‖ GDC-(LiNa)CO$_3$ electrolyte ‖ LSCF (La$_{0.6}$Sr$_{0.4}$Co$_{0.2}$Fe$_{0.8}$O$_{3-\delta}$)-25 wt% (LiNa)CO$_3$ cathode at temperatures ranging from 500-700 °C.

Table: 4. Performance analysis of GDC samples with various amount of (LiNa)CO$_3$ estimated at temperature ranging from 400-800 °C

Temp (°C)	OCV (V)	Current Density (mA cm^{-2})	Power Density (mW cm^{-2})
500	0.8878	540.1371	129.1351
550	0.9317	1356.773	330.0927
600	0.9415	1073.807	266.7095
650	0.9073	2216.838	525.1676
700	0.8829	2225.649	548.5911

4. CONCLUSIONS

20%Gd-CeO$_2$ with various amounts of (LiNa)CO$_3$ have been prepared by mixing 20%Gd-CeO$_2$ nano-powder prepared through chemical synthesis route with (LiNa)CO$_3$. The sintering was performed at sintered at 600 °C for 1 h. The electrical properties have been investigated by using ac impedance spectroscopy. It has been found that bulk resistance of the composites decreases with increasing temperature. The highest ionic conductivity of 0.132 Scm^{-1} (550 °C) and lowest activation energy of 0.42 kJ mol^{-1} in the temperature range of 400-800 °C are shown by the GDC-25 wt% carbonate sample. Addition of LiNaCO$_3$ above 30 wt% increases the bulk resistance. GDC with 25 wt% carbonate composite single cell delivered a peak power density of 550 mW.cm^{-2} at 700 °C.

ACKNOWLEDGEMENT

This work was supported by CSIR, India (Project No: 22 (0524)/10/EMR-II). Authors gratefully acknowledge Indian Institute of Technology Delhi for financial assistance for Ph. D graduate studies, National Physical Laboratory, New Delhi for technical assistance and Department of Science and Technology, New Delhi, India for Travel Grant (No.SR/ITS/4271/2011-2012).

REFERENCES

1. Z. Shao and S. M. Haile, "A High Performance Cathode for the Next Generation Solid-Oxide Fuel Cells," *Nature* 431, 170-173 (2004).

2. B. C. H. Steele and Angelika Heinzel,"Materials for fuel-cell technologies," *Nature* Volume 414, 345-352 (2001).

3. P. Bance , N.P. Brandon , B. Girvan , P. Holbeche, S. O'Dea, B.C.H. Steele, "Spinning-out a fuel cell company from a UK University-2 years of progress at Ceres Power", *J. of Power Sources* 131,86–90, (2004)

4. B. Zhu, S. Li, B. E. Mellander, "Theoretical approach on ceria-based two-phase electrolytes for low temperature (300-600°C) solid oxide fuel cells." Electrochemistry Communications,10, 302-305, (2008)

5. X. Wang, Y. Ma, R. Raza, M. Muhammed and B. Zhu, "Novel core-shell SDC/amorphous Na$_2$CO$_3$ nano composite electrolyte for low temperature SOFCs," Electrochemistry Communications,10, 1617-1620, (2008)

6. J. Huang, Z. Mao, Z. Liu, C. Wang. "Development of novel low-temperature SOFCs with co-ionic conducting SDC-carbonate composite electrolytes." Electrochemistry Communications, 9: 2601-2605, (2007)

7. Rajalekshmi Chockalingam; Jain, S.; Basu, S. "Studies on Conductivity of Composite GdCeO$_2$-Carbonate Electrolytes for Low Temperature Solid Oxide Fuel Cells", *Integrated Ferroelectrics* 115 (2010) 1-12.

8. Rajalekshmi Chockalingam; Basu, S.,"Impedance spectroscopy studies of Gd-CeO$_2$-(LiNa)CO$_3$ nano composite electrolytes for low temperature SOFC applications", *J. Int. Hydrogen Energy*, 36, Issue 22, 14977-14983 (2011).

9. J. Huang, L. Yang, R.Gao, Z. Mao and C. Wang. "A high performance ceramic fuel cell with samarium doped ceria-carbonate composite electrolyte at low-temperatures." Electrochemistry Communications, 8: 785-789, (2006).

10. Rajalekshmi Chockalingam, H. Giesche and V. R. W. Amarakoon "Alumina/Cerium Oxide Nano-Composite Electrolyte for Solid Oxide Fuel Cell Applications", Journal of the European Ceramic Society, 28, pp 959-963, (2008).

11. L. E. Smart and E. A. Moore, "Solid State Chemistry An Introduction" CRC Press, Taylor and Francis Group LLC, Boca Raton, Florida (2005)

12. M. Nauer, C. Ftikos and B. C. H. Steele, "An Evaluation of Ce-Pr Oxides and Ce-Pr-Nb Oxides Mixed Conductors for cathodes of Solid Oxide Fuel cells: Structure, Thermal Expansion and Electrical Conductivity", *J. Euro Ceram Soc.* 14,493-499 (1994).

13. C. M. Lapa, F. M. L. Figueiredo, D. P. F. Desouza, L. Song, B. Zhu, F. M. B. Marques, "Synthesis and characterization of composite electrolytes based on samaria-doped ceria and Na/Li carbonates". *Int. J. of Hydrogen Energy*, 35: 2953-2957, (2010)

14. Z. Tang, Q. Lin, B. E. Mellander, B. Zhu, "SDC-LiNa carbonate composite and nanocomposite electrolytes". *Int. J. of Hydrogen Energy*, 35: 2970-2975,(2010)

15. J. Maier, "Ionic conduction in space charge regions", *Prog. Solid St. Chemistry*, 23: 171-263, (1995)

16. S. P. S. Badwal, H. J. Bruin, A. D. Franklin,"mpedance spectroscopy of the Pt/Yttria doped ceria interface". *Solid State Ionics* 1983; 9 & 10:973-978, (1983)

WEIBULL STRENGTH VARIATIONS BETWEEN ROOM TEMPERATURE AND HIGH TEMPERATURE Ni-3YSZ HALF-CELLS

Declan J Curran[1*], Henrik Lund Frandsen[1], Peter Vang Hendriksen[1]

[1] Mixed Conductors, Department of Energy Conversion and Storage, Technical University of Denmark – DTU, Building 779, P.O. Box 49, 4000 Roskilde, Denmark.
*email: dcur@dtu.dk

ABSTRACT

Solid oxide fuel cell stacks are vulnerable to mechanical failures. One of the most relevant failure mechanisms is brittle fracture of the individual ceramic cells, which are an integral part of the stack structure. Even the mechanical failure of one cell can lead to temporary interruption, reduced efficiency, increased degradation and/or the complete termination of a functioning stack.

This paper investigates the effects of temperature on the mechanical strength of 3% yttria-stabilised zirconia half-cells. Strength was measured using a four-point bend method at room temperature and at 600°C, 700°C and 800°C in a reducing atmosphere. The strength of an as sintered half-cell was also measured at room temperature for comparison. Weibull analysis was performed on large sample sets of 30 for statistical viability. The Weibull strength and elastic modulus of the room temperature tested reduced samples show a decrease of approximately 33% and 51% respectively, when compared to the oxidized samples tested at room temperature. When tested at elevated temperatures both Weibull strength and elastic modulus decrease further when compared to the room temperature reduced samples. However these further reductions are more likely due to thermal interactions with the microstructure of the materials. The Weibull modulus displays a decrease in the room temperature reduced and elevated temperature samples when compared to the room temperature oxidised samples.

1. INTRODUCTION

When in operation solid oxide fuel cell stacks run in the approximate temperature range of 700°C to 850°C depending on design. However increasing cell and stack efficiency is somewhat a false goal if the strength and durability is not advanced in tandem. When running, a stack will operate at approximately 750°C, which can alter the strength characteristics of the individual fuel cells from that at room temperature. This can lead to premature failure of cells that can lead to anything from reduced operation lifetime, reduced efficiency, or total stack failure. Thus, it is imperative to study the mechanical response of individual fuel cells at temperatures in the range of stack operation to understand the material properties.

As ceramic based solid oxide fuel cells are brittle in nature, the failure of an individual cell is difficult to predict as it is rarely preceded by warning signs. In addition, ceramics can fail with large variations in applied stress due to the presence of flaws and varied microstructure, which is very much dependent on the processing steps and handling. The strength of a ceramic can change at elevated temperatures due to thermal effects on the properties of the ceramic, most notably the reduction of the

61

cell. Consequently, testing the strength at ambient temperatures does not give a clear picture of the cell characteristics at elevated temperatures.

In addition to thermal effects within a ceramic that can reduce the overall strength, fuel cells are multilayer components, which brings thermal expansion mismatch into play. The closer the operating temperature of the stack to the temperature at which the cells were assembled the lower these effects. However the assembled temperatures are in excess of 1000°C, thus thermal expansion mismatch is a real concern at stack operating temperatures. These thermal stresses can vary depending on the temperature, the thickness of the layers and the distance from the interfaces between the layers of the cell. Additionally, thermal mismatch also presents a problem when stacks are thermally cycled as residual stresses can build up between the cell layers [1]. Thus, the inclusion of thermal expansion mismatch stresses is paramount to calculating the strength of cells.

The innate nature of ceramics results in a distribution of strength and a common method of failure prediction in ceramics is Weibull statistics [2, 3]. Originally, Weibull published his paper concerning "the applicability of statistics to a wide field of problems" [4] and not specifically to the field of brittle fracture. However this paper applied a method of predicting overall failure due to localised failure making it very applicable to ceramic failure. Eq. 1 shows the Weibull distribution, including volume.

$$P_f(\sigma) = 1 - \exp\left(-\int_v \left(\frac{\sigma}{\sigma_0}\right)^m \frac{dV}{V_0}\right)$$

(1)

$P_f(\sigma)$ is the probability of failure, σ the failure stress, σ_0 the Weibull, or characteristic, strength, m the Weibull modulus, and V_0 the reference volume. The Weibull strength σ_0 is the stress at 63.2% probability of failure and it can be used as an indicator of the overall strength of the material, as a large shift in σ_0 moves the overall strength data to larger strength values and vice versa. The Weibull modulus, m, is related to the flaw size distribution of the material with smaller values indicating the presence of larger, failure inducing, flaws [5].

EXPERIMENTAL

2.1 Materials

Tests were carried out on as-sintered anode supported half-cells. The half-cells consist of an anode support, anode and electrolyte with thicknesses of 300 ± 40 μm, 20 ± 2μm and 10 ± 2 μm, respectively. The cells were produced by tape casting method and lasercut into testable dimensions of 60 mm length by 15 mm width. The cells were randomly sorted into batches of 32 cells for testing under the specific regimes.

2.2 Testing Equipment

Testing was conducted on an in-house built high temperature capable four-point bend machine. This equipment has controlled atmosphere capability, with both hydrogen safety gas (9% H_2 - 91% N_2) and nitrogen (100% N_2) supplies. This allows the samples to be reduced within the testing rig and can allow numerous variations in reduction procedure. Up to 16 samples can be tested in one run. This set up allows the testing of large sets of samples that decreases the probability of experimental error and saves time.

Figure 1: Cells loaded into the rig before a high temperature run.

2.3 Testing Procedure

Five batches of samples were tested: oxidized and reduced samples at room temperature (~22°C), reduced samples at 600°C, reduced samples at 700°C and reduced samples at 800°C. The reduction procedure is as follows:

1. Add reducing atmosphere (9% H_2, 91% N_2) at room temperature.
2. Heat to 500°C at a rate of 1.5°C/min hold for 11 hours (overnight).
3. Ramp to desired temperature at 2.5°C/min and hold for 3 hours at the desired temperature.
4. Mechanically test cells.
5. Drop temperature back to room temperature at 3.0°C/min in reducing atmosphere.
6. Close gases.

For the reduced samples tested at room temperature, procedures 4 and 5 in the above list are reversed.

2.4 Elastic modulus

The elastic modulus (E) was calculated using thin beam theory in a four-point bend context. Using laboratory data the slope of the load-displacement ($\Delta P / \Delta d$) curves for each set of samples was used to determine the elastic modulus by the use of Eq. 2 [6, 7].

$$E = \frac{a(3Lx - 3x^2 - a^2)}{12I} \cdot \frac{\Delta P}{\Delta d} \qquad (2)$$

Where I is the moment of inertia of a rectangular body and x is the position, $a \leq x \leq L/2$ at which the deflection d is measured. d was measured at the position of load application, i.e. $x = a$. The moment of inertia for a rectangular beam equals $wt^3/12$, where w is the width of the beam and t the thickness.

2.5 Statistical analysis

Fracture strength is highly dependent on the distribution of flaws within the material. To quantify this distribution Weibull statistical analysis has been applied to the fracture strength distribution. The probability of failure for a set of measured strengths is obtained by assigning the strength of each specimen in ascending order with a probability of failure. The probability of failure function (P_f), Eq. 3, assigns a probability of failure between 0 and 1 to all of the recorded fracture stresses in the ascending list.

$$P_f = \frac{i - 0.5}{N} \qquad (3)$$

The stress values are ranked in ascending order, $i = 1, 2, 3..., N$, where N represents the total number of tested samples. For the evaluation of the Weibull modulus, Eq. 1 is rearranged to give Eq. 4, which results in a straight line fit where the Weibull modulus can be determined by obtaining the slope of the straight line [8]. This is done for ease of access of information, as the straight line plot allows easy determination of the slope of the data [9].

$$\ln\left[\ln\left(\frac{1}{1 - P_f}\right)\right] = m \ln \sigma - m \ln \sigma_0 \qquad (4)$$

To achieve a low uncertainty, 30 specimens for each testing regime were used in the analytical evaluation of the Weibull distribution [10].

2.6 Thermal stress analysis

In addition to stresses, which arise from the application of an external load, thermal expansion mismatch between layers can also elicit thermal stresses within a sample. The thermal stresses were included in the stress analysis by classical laminate theory, see e.g. ref [11]. Here it is applied to a 4 point bending case.

Note: this study only investigates the normal stresses along the beam axis and does not consider shear stresses.

RESULTS & DISCUSSION
Figure 2 below shows the Weibull plots of the sample sets tested over the temperature range of 22°C to 800°C. The changes in Weibull strength and elastic modulus are detailed over these temperatures.

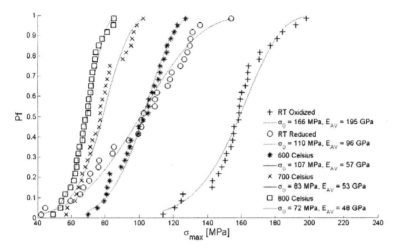

Figure 2: Weibull plots of probability of failure vs. max stress at fracture for room temperature (RT) and elevated temperatures.

The room temperature oxidized sample set displays the largest Weibull strength with values of 166 MPa, see Figure 2. These values decrease by approximately 33% when considering the reduced samples tested at room temperature (RT reduced). As these sample sets are tested at the same temperature, it was concluded that the drop in Weibull strength between these sample sets is mainly due to the reduction process. This has been shown in a number of studies [12, 13, 14, 15] since one of the main effects of the reduction process is an increase in porosity, which occurs as the larger nickel oxide (NiO) is converted to the smaller nickel (Ni) particles. This can have a defining control over the Weibull strength of a porous material with an exponential relationship, see Eq. 5. Where σ_{0p} is zero-porosity Weibull strength, σ_0 is the Weibull strength at volume fraction porosity p, and b is a material constant.

$$\sigma_0 = \sigma_{op} \exp(-bp) \tag{5}$$

The change in porosity is also the main cause of the 51% drop in elastic modulus, shown in Table (i), between the room temperature oxidized and the room temperature reduced samples. When considering materials with large amounts of porosity the minimum solid area exponential relationship, Eq.5, may be used to describe the change in E with porosity [16]. Although further investigations would be necessary to verify this.

Table (i) below details the change in average elastic modulus (E_{AV}), Weibull strength σ_0 and the normalised σ_0 across the testing temperature.

Temperature (°C)	Elastic Modulus (E_{AV}) (GPa)	Weibull Strength (σ_0) (MPa)	σ_0^2/E
22 (oxidized)	194	166	227
22 (reduced)	95	110	84
600	56	107	30
700	52	83	34
800	48	72	32

At the high testing temperatures of 600°C, 700°C and 800°C there is a strong decrease in both Weibull strength and elastic modulus with increasing temperature from room temperature, see Table (i). As these samples are reduced to the same degree as the room temperature reduced samples it is clear that the high temperatures are having an effect on the Weibull strength and elastic modulus, in addition to the negative effects due to the reduction process. The decrease in both of these parameters is likely linked to the increasing atomic distance that occurs between atoms of the samples with increasing thermal energy. According to Griffith, the fracture strength σ_f of a material is dependent on a number of properties, see Eq. 6

$$\sigma_f = \alpha \left(\frac{2E\gamma_s}{c} \right)^{1/2}$$

(6)

Where α is a numerical factor depending on the crack geometry, γ_s is the surface energy and C is the flaw length. At room temperature these values are considered constant, however they will change with temperature making fracture strength dependent on temperature in an indirect manner. As temperature increases the interatomic distance increases, increasing the unit size of the material and reducing the bonding strength between the atoms. The reduction of the interatomic bonding strength reduces the bond stiffness between those atoms. On a macro scale this reduces the elastic modulus as the overall bonding strength between the atoms of the material decreases making it more elastic.

However this is only a part of the explanation, which can be realised from the final column in Table (i). The strength has been normalised with the elastic modulus, σ_0^2/E, according to Eq. (6). The term σ_0^2/E is proportional to surface energy with the only constant being the proportionality factor. Therefore, the surface energy must also change with temperature, although it seems relatively constant at the higher temperatures. This decrease in bonding strength between adjacent atoms and the reduction in surface energy (γ_s) of the material reduces the energy needed to cleave the atoms apart at the crack tip to create two new surfaces.

Table (ii): Weibull modulus (m) changes at room temperature and elevated temperatures.

Temperature (°C)	Weibull Modulus (m)
22 (oxidized)	9.0
22 (reduced)	4.1
600	7.9
700	6.7
800	8.7

The room temperature reduced samples experience the largest drop in Weibull modulus, with an approximate drop of 54% to a value of 4.1 when compared to the room temperature oxidized samples which display a Weibull modulus of 9.0, see Table (ii). The sample sets tested at the high temperatures do not experience such a large reduction in Weibull modulus as compared to the room temperature reduced samples, with the lowest value of 6.7 being displayed by the 700°C samples. It must be noted that the Weibull values for the high temperatures sample sets are within the error limits (~25%) and as such will be considered the same for this discussion.

When reduction occurs there may be an increase in the flaw distribution by the generation of pores, or flaws, as a result of the reduction process as the NiO decreases in size while converting to Ni [12]. However this explanation does not explain the 46% difference in Weibull modulus between the average of the reduced high temperature sample sets and the room temperature reduced samples. It is likely that the cooling process, which is relatively rapid compared to the reduction process, introduces flaws in the materials so as to further increase the flaw distribution giving a lower Weibull modulus. In addition to the flaws being introduced by relatively rapid cooling, it is also feasible that the residual stresses [17] may also play a role in the large fracture distribution as well as influencing the fracture strength. However for a firm conclusion on this hypothesis, further experiments would need to be conducted.

CONCLUSION

When operating half-cells at elevated temperatures it is shown that the elastic modulus decreases with increasing temperature. The largest difference in elastic modulus, 51%, is experienced between samples tested at room temperature in an oxidized state and samples tested at room temperature in a reduced state. This implies that the reduction process has the single greatest affect on the elastic modulus. Testing at high temperatures decreased the elastic modulus even further, although this decline was not as severe as the original decline due to reduction.

The decrease in elastic modulus with increasing temperature was matched by a decrease in the Weibull strength. At room temperature the drop in Weibull strength was mainly due to the reduction process and an increase in the porosity of the cells. At high temperature the incremental decrease in Weibull strength below that of the room temperature reduced samples was due to the increasing atomic distance, the decreasing of elastic modulus and the subsequent decrease in the surface energy. This resulted in less energy needed to fracture the samples as less energy was needed to create two new fracture surfaces ahead of the crack tip.

The reduction in Weibull modulus in the reduced samples is due to the increase in porosity of cells which results from the conversion of nickel oxide (NiO) to nickel (Ni). The room temperature reduced samples show the lowest values of Weibull modulus. These samples are reduced in the same manner as the samples tested at high temperature, which display larger Weibull modulus values. This suggests that the cooling process may be directly damaging the microstructure and giving an increased flaw distribution in the cooled samples, which has serious implications for thermal cycling.

REFERENCES

1. B. Sun, R.A. Rudkin and A. Atkinson. "Effect of Thermal Cycling on Residual Stress and Curvature of Anode-Supported SOFCs," Fuel Cells, **9,** 805-13 (2009).

2. N. Ramakrishnan and V.S. Arunachalam. "Effective elastic moduli of porous solids," Journal of Materials Science, **25,** 3930-7 (1990).

3. J. Malzbender and R.W. Steinbrech. "Fracture test of thin sheet electrolytes for solid oxide fuel cells," Journal of the European Ceramic Society, 2597 (2007).

4. W. Weibull, "A statistical distribution function of wide applicability," Journal of Applied Mechanics, 293 (1951).

5. A.D.S. Jayatilaka and K. Trustrum. "Statistical approach to brittle fracture," Journal of Materials Science, **12,** 1426-30 (1977).

6. S.P. Timoshenko, *Mechanics of Materials;* VanNostrand Reinhold, New York, 1972.

7. J. Malzbender and R.W. Steinbrech. "Mechanical properties of coated materials and multi-layered composites determined using bending methods," Surface and Coatings Technology, **176,** 165-72 (2004).

8. F.L. Lowrie and R.D. Rawlings. "Room and high temperature failure mechanisms in solid oxide fuel cell electrolytes," Journal of the European Ceramic Society, **20,** 751-60 (2000).

9. J.B. Quinn and G.D. Quinn. "A practical and systematic review of Weibull statistics for reporting strengths of dental materials," Dental Materials, **26,** 135-47 (2010).

10. A. Khalili and K. Kromp. "Statistical properties of Weibull estimators," Journal of Materials Science, **26,** 6741-52 (1991).

11. H.L. Frandsen, T. Ramos, A. Faes, M. Pihlatie and K. Brodersen. "Optimization of the strength of SOFC anode supports," Journal of the European Ceramic Society, (2011).

12. M. Radovic and E. Lara-Curzio. "Mechanical properties of tape cast nickel-based anode materials for solid oxide fuel cells before and after reduction in hydrogen," Acta Materialia, 5747 (2004).

13. M. Radovic and E. Lara-Curzio. "Elastic Properties of Nickel-Based Anodes for Solid Oxide Fuel Cells as a Function of the Fraction of Reduced NiO," Journal of the American Ceramic Society, 2242 (2004).

14. Y. Wang, M.E. Walter, K. Sabolsky and M.M. Seabaugh. "Effects of powder sizes and reduction parameters on the strength of Ni–YSZ anodes," Solid State Ionics, **177,** 1517-27 (2006).

15. M. Pihlatie, A. Kaiser and M. Mogensen. "Mechanical properties of NiO/Ni–YSZ composites depending on temperature, porosity and redox cycling," Journal of the European Ceramic Society, 1657 (2009).

16. R.W. Rice, "Evaluation and extension of physical property-porosity models based on minimum solid area," Journal of Materials Science, **31,** 102-18 (1996).

17. C.H. Hsueh, "Thermal stresses in elastic multilayer systems," Thin Solid Films, 182 (2002).

IN-SITU XRD OF OPERATING LSFC CATHODES: DEVELOPMENT OF A NEW ANALYTICAL CAPABILITY

John S. Hardy, Jared W. Templeton, and Jeffry W. Stevenson
Pacific Northwest National Laboratory
Richland, WA, USA

ABSTRACT

A solid oxide fuel cell (SOFC) research capability has been developed that facilitates measuring the electrochemical performance of an operating SOFC while simultaneously performing x-ray diffraction on its cathode. The evolution of this research tool's development is discussed together with a description of the instrumentation used for in-situ x-ray diffraction (XRD) measurements of operating SOFC cathodes. The challenges that were overcome in the process of developing this capability, which included seals and cathode current collectors, are described together with the solutions that are presently being applied to mitigate them.

INTRODUCTION

Cathodes are often regarded as the limiting factor in anode-supported SOFC performance.[1-5] It is for this reason that much of the research effort in SOFCs is directed at developing a better understanding of cathodes, including the materials and the mechanisms occurring during cell operation. Thus, any new means of characterization, analysis, or data collection that can be applied to obtain useful information regarding the cathode is of great value in adding to the compendium of knowledge which is continually leading toward a better understanding of these materials. In this paper, we will discuss the development of one such characterization technique that has only recently begun to be employed in cathode research.

One mechanism or set of mechanisms that is yet little understood is the one governing the degradation in SOFC performance that is attributable to the cathode. Chromium poisoning[6-16], strontium segregation[13-21], humidity in the cathode gas[6; 17; 18], silicon poisoning[18], and gadolinium diffusion from the barrier layer[21] have all been investigated for their role in degradation. However, it has been difficult to pinpoint the dominant mechanism(s) responsible for performance degradation during cell operation definitively enough for a general consensus to be reached amongst the SOFC research community. It is hoped that the research tool discussed here can be employed to shed additional light on the source of cathode degradation.

The benefit of performing XRD measurements on the cathode while it operates is that any changes detectable by XRD can be monitored as a function of time while the cell operates as opposed to searching for changes in the cathode after the cell test is discontinued and disassembled. This avoids failing to observe reversible changes or not fully recognizing the magnitude of those that are partially reversible. In best case scenarios, it may make it possible to correlate the extent of detected changes to the degree of degradation in cell performance. It also makes it possible to directly measure the effects of changes in experimental parameters on the cathode and correlate these to cell performance. For example, operating voltage, temperature, and oxygen content can be varied or a contaminant can be introduced.

XRD is a powerful analysis tool in that it can monitor the phase composition of the cathode based on the locations of diffraction peaks. Therefore, if a new phase is formed in sufficient quantity that new peaks arise in the XRD pattern, an effort can be made to identify it. Phase transitions can also be monitored. For example, a transition from cubic to rhombohedral

71

symmetry would be distinguished by peak splitting. Strain in the crystal lattice is manifest by peak shifts. Changes in crystallite size are revealed by broadening or narrowing of the diffraction peaks. Therefore, XRD of an operating cathode makes it possible to monitor all of these aspects of the cathode as a function of time or other experimental parameters as described above.

EXPERIMENTAL

Cell Preparation

The 8% yttria-stabilized zirconia (8YSZ) electrolyte layer, the active anode layer, and the bulk anode layer were separately tape cast and subsequently laminated together. Discs were then cut from the laminate and sintered at 1385°C for 2 hours in air to create anode-supported electrolyte substrates that were approximately 13 mm in diameter and 1 mm thick including approximately 8 μm thick dense electrolyte membranes. Samaria-doped ceria (SDC) interlayers were screen printed on the anode-supported 8YSZ membranes using an ink made from 40 wt% SDC powder [Praxair Specialty Ceramics, Woodinville, WA] with an average particle size of approximately 0.2 μm in V-006 binder [Heraeus, West Conshohocken, PA]. The SDC interlayers were sintered at 1200°C for 2 hours in air together with the Ni mesh embedded in NiO paste that was used as the anode current collector. $La_{0.6}Sr_{0.4}Co_{0.2}Fe_{0.8}O_{3-\delta}$ (LSCF) cathode powder [Praxair Specialty Ceramics] was attrition milled for 5-10 hours to reduce the average particle size to approximately 0.35 μm. It was then mixed into V-006 binder at 40 wt% solids loading and screen printed on the SDC interlayer and sintered at 1100°C for 2 hours in air. The sintered cathode was circular and had an area of 0.5 cm^2, which was the active cell area used in calculating power densities. Various cathode current collector configurations were assessed for adequate current collection and minimal interference with the x-rays meant to impinge on the LSCF cathode. Sealing procedures were also evaluated for bonding the cells to the alumina test fixture described below. The test results from cells using the current collectors and seals that were evaluated will be discussed hereafter.

XRD-compatible SOFC Test Fixture

A button cell test fixture was designed to operate small-scale button cells in a high temperature x-ray diffractometer (HTXRD) while collecting diffraction spectra from the working cathode. As can be seen in Figure 1, the XRD test fixture is very similar to a typical button cell test fixture, except that it is scaled down in size and is coupled to the sample holder flange used for inserting XRD specimens into the heating chamber. A four bore 0.25 inch diameter alumina tube runs up the center of the outer ½ inch diameter alumina tube. The bores in the inner tube provide a pathway for platinum contact leads and humidified hydrogen fuel gas to reach the anode. Meanwhile, the platinum leads for the cathode current collector run down the outside of the ½ inch alumina tube and out of the XRD chamber.

Figure 1. The original button cell test fixture constructed for the electrochemical tests of cells during XRD analysis.

Electrochemical Cell Testing

The cell test fixture was inserted into the HTK 1200 heating chamber [Anton Paar, Ashland, VA] of the D8 Advance XRD [Bruker AXS, Madison, WI] shown in Figure 2. In keeping with the heating schedule typically used for SOFCs, which often includes a sealing step, the cell was heated to 830°C for 1 h as is done for glass sealing. It was then cooled to the 750°C operating temperature. Hydrogen with 3% water vapor flowing at a rate of 75 sccm was then introduced to the anode which was reduced from NiO to Ni metal before adjusting the operating voltage of the cell. After reduction, the composition of the functional anode layer was 50 vol% Ni and 50 vol% 8YSZ and the bulk anode layer was 40 vol% Ni and 60 vol% 8YSZ. Current-voltage and electrochemical impedance spectroscopy (EIS) data were collected using a Solartron 1480 Multistat and 1255 Frequency Response Analyzer. Performance data was recorded under constant current conditions at the current initially measured at an operating voltage of 800 mV. During intermittent EIS measurements, cells were subjected to an ac amplitude of 20 mA. Air was supplied to the cathode at 350 sccm.

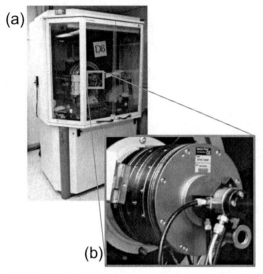

Figure 2. (a) The Bruker D8 Advance XRD with (b) an Anton Paar HTK 1200 heating chamber used for in-situ XRD measurements of operating SOFCs.

X-ray Diffraction

Over the duration of cell testing, ten minute scans were repeated continuously over an angular range of 25 to 85° 2θ with 0.05° steps and a 0.43 second count time per step. The XRD was outfitted with a Cu $K_{\alpha1}$ radiation source, a Göbel mirror, a 0.12° diffracted beam Soller slit, and a Sol-X energy dispersive x-ray detector [Bruker AXS] that filters out K_β peaks and the fluorescence background that arises due to the presence of Co and Fe.

RESULTS AND DISCUSSION

A number of challenges were encountered in the initial attempts to operate a SOFC while performing XRD on its cathode. This was not surprising since there was no precedence for such measurements found in the literature. The main challenges were related to sealing and cathode current collection. Sealing was difficult because the manufacturer advised against burning out organics in the heating chamber. This implied that the binder in the seal material would have to be burned out in another furnace prior to testing in the XRD. Additionally, sealing often requires a compressive load to be applied while the seal is fired. However, XRD requires x-rays to impinge on the specimen, therefore the means of applying the load must not disrupt the path of the x-rays. The load must either be removed between completion of sealing and initiation of XRD scans or be applied in such a way as not to interfere with the x-ray beam during XRD. At first glance, the obvious solution would appear to be sealing the cell to the test fixture in a separate furnace where a load can be applied, and subsequently removed, before transporting the assembly to the XRD. However, due to the thermal expansion mismatch between the SOFC and the alumina test fixture, cooling from the sealing temperature to room temperature when constrained by a rigid seal often causes the cell to fracture. Therefore, other sealing procedures were explored as described below.

The two seal materials with which the authors had the most experience in SOFC testing are Ceramabond 569 & 685 [Aremco, Valley Cottage, NY] and G-18 barium calcium aluminosilicate glass (developed at PNNL). A comparison of the two materials based on the experience of the authors is given in Table I. The Ceramabond seal has several apparent advantages over G-18 for tests in the XRD in that the organics can be removed with minimal heat treatment and immediately thereafter the bond is strong enough for handling. Additionally, a seal can be reliably formed without applying a compressive load. However, the authors typically observe open circuit voltages (OCVs) of only ~1.06 V when using Ceramabond seals, whereas cells with G-18 seals exhibit OCVs of ~1.10 V. Unfortunately, however, G-18 requires a much higher temperature to burn out the organics and is very weakly bonded thereafter. Moreover, sealing without a compressive load is inconsistent.

Table I. Comparison of two candidate sealing materials.

Aremco Ceramabond 569 & 685	PNNL G-18 Barium Calcium Aluminosilicate Glass
Organics removed during 100C curing heat treatment	Organics must be burned out at 500-600C
Bond is strong enough for handling after curing	Not much strength for handling after burn out
Seals reliably without a load	Hit or miss on sealing without a load
Not a great seal (typical OCV ~1.04 V)	Good seal (typical OCV ~1.1 V)

As is the case for applying a load during firing, cathode current collection is also problematic because of the necessity for the x-rays to reach the cathode surface. Therefore, a balance must be achieved in which the current collector has adequate coverage of the cathode and minimal interaction with the impinging x-ray beam. Thus, various current collector configurations were tested.

The first attempt at conducting in-situ XRD measurements of the cathode utilized a Ceramabond seal and a fine gold mesh (100 mesh, 0.064mm dia. wire) that was rolled to a thickness of less than 40 μm for the cathode current collector. The OCV was 1.04 V, indicating seals that were less than perfect but adequate for cell operation. The electrochemical performance of the cell is plotted as a function of time in Figure 3. The inset shows EIS spectra that were collected intermittently. This initial test was encouraging in that it proved the feasibility of operating a cell for over 150 h in the heating chamber of the XRD using the newly designed test fixture. From the EIS results, it can be seen that the rapid degradation in performance was due to monotonic increases in both the ohmic and polarization resistance of the cell. It was also encouraging that XRD patterns were simultaneously measured from the cell although, unfortunately, only the gold cathode current collector and the Ceramabond seal material, which forms a ridge atop the circumference of the cell were detected. Apparently, the LSCF was completely shielded from the incident x-ray beam because no LSCF peaks were detected. This can be seen in the XRD pattern in Figure 4 that is representative of the typical pattern collected during cell operation.

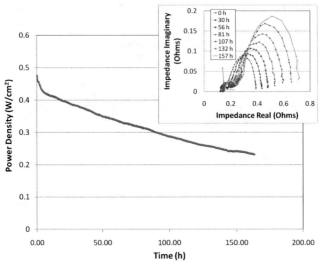

Figure 3. Results of electrochemical tests of the first cell operated in the XRD test fixture.

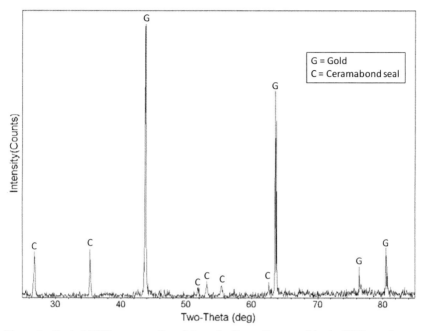

Figure 4. Typical XRD pattern collected from the first cell operated in the XRD test fixture, which utilized a Ceramabond seal and a fine Au mesh current collector that had been rolled to a thickness of less than 40 μm.

Based on the results of the first cell, adjustments were made when preparing the second cell in an effort to allow the x-rays to reach the LSFC cathode. The height of the ridge formed by the Ceramabond seal was minimized and a thin layer of gold was screen printed on the cathode surface for current collection. However, an XRD pattern taken at 800°C (see Figure 5) found that while x-ray detection of the Ceramabond seal was greatly reduced, the thin screen printed gold layer was still impenetrable by the x-rays, leaving peaks from the underlying LSCF still undetected. The decreased height of the Ceramabond seal at the edge of the cell is evident in the fact that the gold peak at ~38° 2θ was no longer shielded from detection as it had been on the first cell (see Figure 4), rather the incident x-rays at an angle of ~19° from the plane of the SOFC were high enough to be above the ridge of Ceramabond. Upon finding that the cathode was not detected by XRD, it was decided to forego electrochemical measurements on this cell.

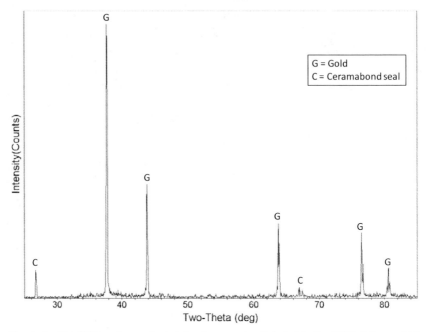

Figure 5. The XRD pattern collected for the second cell tested in the XRD fixture, which utilized a Ceramabond seal with reduced ridge height and a thin screen printed Au paste current collector.

Table 2 is a summary of a series of subsequent tests which suffered from inadequate sealing. The third cell prepared for XRD testing utilized the same seal as the second cell. In order to ensure that the x-rays would not be shielded from impinging on the cathode, a gold mesh ring was bonded to the perimeter of the cathode using gold paste dots as contacts, leaving the center of the cathode open for XRD. With this current collector configuration, the main concern was whether it would provide adequate current collection. However, measurement of the OCV revealed that the seal was not gas tight. Upon finding that the Ceramabond seal was not infallible under the sealing conditions required for the XRD cell tests, the fourth cell attempted to incorporate a G-18 seal that was applied around the edge of the cell and tube, while again using the perimeter mesh ring for current collection. Upon finding that sealing had failed, the cell was cooled and an examination of the seal discovered that the G-18 had melted and run down the tube. This led to an investigation comparing the actual temperature inside the heating chamber to the temperature measured by the controller. It was found that the temperature at the location of the SOFC was ~50°C higher than the controller temperature. Thus, corrections for this discrepancy were made in all subsequent tests. The next cell utilized G-18 paste around the edge of the cell and tube. The G-18 was then coated with Ceramabond, but this seal still was not adequate. The following cell used a ring of G-18 tape between the top of the tube and the bottom of the cell in addition to the configuration used in the previous test and provided an OCV of ~1.01 V. This would have been adequate to operate the cell, except that the OCV was not stable and would intermittently drop to significantly lower voltages. For the seventh cell, the G-18 tape and paste were used again, as had been done in the previous test, except without the Ceramabond coating. The OCV was not measured because no bubbling was observed in the exhaust line bubbler. Upon inspection, it was found that water had been condensing in the vent line that led up into the building exhaust and eventually created enough back pressure to cause difficulty in obtaining a quality seal. Therefore, the vent line was cleared and a moisture trap was installed to keep the line clear.

Table II. Tests in the XRD fixture that suffered from inadequate sealing.

Test:	Seal Material:	Current Collector:	OCV:	Findings:
3	Ceramabond	Gold mesh ring around perimeter of cathode with gold paste contact points	0.850 V	
4	G-18 Paste around edge of cell and tube	Gold mesh ring around perimeter of cathode with gold paste contact points	0.07 V, G-18 melted and ran down tube	The cell had been 50°C hotter than expected
5	G-18 Paste with Ceramabond painted over G-18	Gold mesh ring around perimeter of cathode with gold paste contact points	0.60 V	
6	G-18 Paste with Ceramabond painted over it and G-18 tape between cell and tube	Gold mesh ring around perimeter of cathode with gold paste contact points	1.01 V but not stable	
7	G-18 Paste and G-18 tape	Gold mesh ring around perimeter of cathode with gold paste contact points	Not Measured – No bubbling in exhaust line	Water was condensing & plugging the vent

It was not until the eighth cell, which returned to utilizing the Ceramabond seal while again endeavoring to collect current using a gold mesh ring around the perimeter of the cathode, that LSCF peaks were successfully detected in XRD spectra collected during cell operation. The results of this test are reported elsewhere.[22]

Since conducting these experiments, additional improvements to the in-situ XRD tests have been made. An air delivery tube has been attached to the heating chamber as shown in Figure 6 to direct the incoming cathode air to flow directly onto the surface of the cathode. In the prior tests discussed above, the cathode air was introduced through one port in the heating chamber and exited through another with no guarantee that there was sufficient air flow in the vicinity of the cathode surface. Additionally, the slit that controls the width of the x-ray beam that irradiates the SOFC needed adjustment. The length of the slit, shown in the top of the picture in Figure 7, is long enough that it allows the beam to impinge on the ring of gold mesh covering the perimeter of the cathode. In order to eliminate as much diffraction data arising from materials other than the cathode as possible, the insert at the bottom of Figure 7 was fabricated to shorten the beam and confine it to the uncovered portion of the cathode.

Figure 6. An air delivery tube that directs incoming cathode air directly at the surface of the cathode.

Figure 7. The insert at the bottom of the picture was fabricated for incorporation into the XRD slit assembly above. While the slit assembly narrows the incident x-ray beam, the insert shortens the length of the beam so that it impinges only on the center of the cathode.

The cell test fixture has also undergone a few enhancements since completion of the tests mentioned above. The enhanced test fixture is depicted in Figure 8. Four alumina rods surround the outer alumina tube and transmit a compressive load to the cell during sealing from springs that are located at their base, outside of the hot zone. The spring loaded rods pull down on a thin alumina washer placed atop the outer perimeter of the fuel cell. The inside diameter of the washer is large enough to allow the x-ray beams to pass to and from the cathode surface without interference. This has now made it possible to use G-18 seals, resulting in OCVs of ~1.10 V. An S type thermocouple has also been incorporated for measuring the temperature of the cell. The thermocouple wires extend up the test fixture along the outside of the ½ inch alumina tube and thin alumina washer and terminate in the junction that is within 1.0 mm above the cathode surface.

Figure 8. The XRD cell test fixture with enhancements that have been added since the completion of the tests mentioned above.

CONCLUSION

A new tool for the characterization of SOFC cathodes has been developed and is currently being applied toward the goal of increasing the understanding of SOFC cathodes. Along the path of development of the capability for in-situ XRD measurements of operating SOFC cathodes, a number of challenges were discovered and overcome sufficiently that simultaneous XRD and electrochemical measurements could be performed on anode-supported cells. The two main challenges included sealing and cathode current collection. After a number of trials that were not entirely successful due to the cathode being shielded from detection by XRD or inadequate sealing, the first successful test utilized a seal consisting of Ceramabond 569 & 685 and a cathode current collector consisting of a ring of gold mesh that contacted the cathode with a ring of gold paste dots. Additional improvements have since been incorporated, including improved cathode air delivery, better targeting of the cathode with the incident x-ray beam, and an enhanced XRD-compatible cell test fixture that makes it possible to seal with G-18 glass through spring loading and monitor the temperature of the cell with a thermocouple within 1 mm of the cathode surface. There are still improvements to be made, such as increasing the measured power density, but much progress has been made in successfully performing in-situ XRD measurements of anode-supported cells and further improvements are still being pursued.

REFERENCES

[1] F. Bidrawn, R. Kungas, J. M. Vohs, andR. J. Gorte, "Modeling Impedance Response of SOFC Cathodes Prepared by Infiltration," *Journal of the Electrochemical Society,* **158**[5] B514-B25 (2011).

[2] J. M. Vohs and R. J. Gorte, "High-Performance SOFC Cathodes Prepared by Infiltration," *Adv. Mater.,* **21**[9] 943-56 (2009).

[3] F. Tietz, Q. Fu, V. A. C. Haanappel, A. Mai, N. H. Menzler, andS. Uhlenbruck, "Materials development for advanced planar solid oxide fuel cells," *Int. J. Appl. Ceram. Technol.,* **4**[5] 436-45 (2007).

[4] S. B. Adler, "Factors governing oxygen reduction in solid oxide fuel cell cathodes," *Chem. Rev.,* **104**[10] 4791-843 (2004).

[5] S. J. Skinner, "Recent advances in Perovskite-type materials for solid oxide fuel cell cathodes," *Int. J. Inorg. Mater.,* **3**[2] 113-21 (2001).

[6] J. J. Bentzen, J. V. T. Hogh, R. Barfod, andA. Hagen, "Chromium Poisoning of LSM/YSZ and LSCF/CGO Composite Cathodes," *Fuel Cells,* **9**[6] 823-32 (2009).

[7] M. C. Tucker, H. Kurokawa, C. P. Jacobson, L. C. De Jonghe, andS. J. Visco, "A fundamental study of chromium deposition on solid oxide fuel cell cathode materials," *Journal of Power Sources,* **160**[1] 130-38 (2006).

[8] S. P. Simner, M. D. Anderson, G. G. Xia, Z. Yang, L. R. Pederson, andJ. W. Stevenson, "SOFC performance with Fe-Cr-Mn alloy interconnect," *Journal of the Electrochemical Society,* **152**[4] A740-A45 (2005).

[9] E. Konysheva, H. Penkalla, E. Wessel, J. Mertens, U. Seeling, L. Singheiser, andK. Hilpert, "Chromium poisoning of perovskite cathodes by the ODS alloy Cr5Fe1Y(2)O(3) and the high chromium ferritic steel Crofer22APU," *Journal of the Electrochemical Society,* **153**[4] A765-A73 (2006).

[10] J. W. Fergus, "Effect of cathode and electrolyte transport properties on chromium poisoning in solid oxide fuel cells," *International Journal of Hydrogen Energy,* **32**[16] 3664-71 (2007).

[11] J. Guan, S. Zecevic, Y. Liu, P. Lam, R. Klug, M. Alinger, S. Taylor, B. Ramamurthi, R. Sarrafi-Nour, andS. Renou, "Performance Degradation of Solid Oxide Fuel Cells with Metallic Interconnects," pp. 405-12. in Solid Oxide Fuel Cells 10, **Vol. 7.** *ECS Transactions.* Edited by K. Eguchi, S. C. Singhai, H. Yokokawa, andH. Mizusaki, 2007.

[12] J. Y. Kim, V. L. Sprenkle, N. L. Canfield, K. D. Meinhardt, andL. A. Chick, "Effects of chrome contamination on the performance of La0.6Sr0.4Co0.8O3 cathode used in solid oxide fuel cells," *Journal of the Electrochemical Society,* **153**[5] A880-A86 (2006).

[13] M. R. Ardigo, A. Perron, L. Combemale, O. Heintz, G. Caboche, andS. Chevalier, "Interface reactivity study between La(0 6)Sr(0 4)Co(0 2)Fe(0 8)O(3-delta) (LSCF) cathode material and metallic interconnect for fuel cell," *Journal of Power Sources,* **196**[4] 2037-45 (2011).

[14] X. B. Chen, L. Zhang, andS. P. Jiang, "Chromium deposition and poisoning on (La0.6Sr0.4-xBax)(Co0.2Fe0.8)O-3 (0 <= x <= 0.4) cathodes of solid oxide fuel cells," *Journal of the Electrochemical Society,* **155**[11] B1093-B101 (2008).

[15] C. W. Sun, R. Hui, andJ. Roller, "Cathode materials for solid oxide fuel cells: a review," *Journal of Solid State Electrochemistry,* **14**[7] 1125-44 (2010).

[16] H. Yokokawa, H. Y. Tu, B. Iwanschitz, andA. Mai, "Fundamental mechanisms limiting solid oxide fuel cell durability," *Journal of Power Sources,* **182**[2] 400-12 (2008).

[17]R. R. Liu, S. H. Kim, S. Taniguchi, T. Oshima, Y. Shiratori, K. Ito, andK. Sasaki, "Influence of water vapor on long-term performance and accelerated degradation of solid oxide fuel cell cathodes," *Journal of Power Sources,* **196**[17] 7090-96 (2011).

[18]E. Bucher and W. Sitte, "Long-term stability of the oxygen exchange properties of (La,Sr)(1-z)(Co,Fe)O(3-delta) in dry and wet atmospheres," *Solid State Ionics,* **192**[1] 480-82 (2011).

[19]C. Endler, A. Leonide, A. Weber, F. Tietz, andE. Ivers-Tiffee, "Time-Dependent Electrode Performance Changes in Intermediate Temperature Solid Oxide Fuel Cells," *Journal of the Electrochemical Society,* **157**[2] B292-B98 (2010).

[20]S. P. Simner, M. D. Anderson, M. H. Engelhard, andJ. W. Stevenson, "Degradation mechanisms of La-Sr-Co-Fe-O3SOFC cathodes," *Electrochem. Solid State Lett.,* **9**[10] A478-A81 (2006).

[21]S. Uhlenbruck, T. Moskalewicz, N. Jordan, H. J. Penkalla, andH. P. Buchkremer, "Element interdiffusion at electrolyte-cathode interfaces in ceramic high-temperature fuel cells," *Solid State Ionics,* **180**[4-5] 418-23 (2009).

[22]J. S. Hardy, J. W. Templeton, D. J. Edwards, Z. Lu, andJ. W. Stevenson, "Lattice expansion of LSCF-6428 cathodes measured by in situ XRD during SOFC operation," *Journal of Power Sources,* **198**[0] 76-82 (2012).

PROTON CONDUCTION BEHAVIORS IN Ba- AND Mg-DOPED LaGaO₃

Xuan Zhao, Nansheng Xu, Kevin Romito, Kevin Huang*
Department of Mechanical Engineering, University of South Carolina, Columbia, SC29201, USA

Alisha Lucas, Changyong Qin
Department of Chemistry, Benedict College, Columbia, SC29205, USA

ABSTRACT
Proton conducting oxides have potential application as an electrolyte for intermediate temperature solid oxide fuel cells. This paper reports the proton conduction behavior of a Ba- and Mg-doped LaGaO₃ ($La_{1-x}Ba_xGa_{0.80}Mg_{0.20}O_{3-\delta}$, LBGM). XRD, SEM and AC impedance spectroscopy are employed to characterize the synthesized LBGM. SEM reveals a dense microstructure at x=0.10 and XRD indicates it as a single phase. At a fixed water partial pressure of 0.065 atm, LBGM with x=0.10 also exhibits the highest conductivity. It is also found that the grain boundary conductivity shows a strong dependence on water vapor partial pressure in oxidizing atmospheres, suggesting its proton conducting nature. In contrast, the bulk conductivity shows less dependence on water vapor partial pressure in reducing atmosphere, inferring dominant oxide-ion conduction.

INTRODUCTION

Proton conducting oxides are promising electrolyte materials for intermediate temperature solid oxide fuel cells (ITSOFCs), hydrogen separation reactors, and gas sensors. Among these proton conductors, Ba-containing perovskites such as doped BaZrO₃ and BaCeO₃ are most studied and exhibit the highest proton conductivity [1-3]. One of the important characteristics of good proton conducting perovskite oxides, as supported by both experiments and modeling, is the high symmetry of crystal structure – cubic, which presents the lowest energy barrier to the migration of protons. However, this class of materials either experiences the difficulty of achieving dense microstructure or inferior chemical stability. Therefore, there is an incentive to develop alternative proton conductors.

Sr- and Mg-doped LaGaO₃ (LSGM) is one family of perovskite-structured fast-ion conductors discovered in 1990s [4]; although LSGM was discovered mainly as an oxide-ion conductor, Huang *et al* [5] first pointed out proton conduction in LSGM, which was subsequently verified by Ma *et al* [6]. Recently, proton conductivity was also observed in LaBaGaO₄ [7], a structure belonging to K₂NiF₄ type and containing tetrahedral moieties GaO₄. By varying the self-doping content of Ba within the series $La_{1-x}Ba_{1+x}GaO_4$, the highest proton conductivity was found at x=0.20 [7].

LaBaGaO₄ has a non-cubic structure [8], which presents an unfavorable environment for proton migration and therefore may limit the level of proton conductivity. To retain the favorable cubic structure, we wonder if proton conductivity can be improved by replacing Sr in LSGM with Ba. Ba has a larger ionic radius than Sr, which renders the ability that coordinates with more oxygen, facilitating the water incorporation process into oxygen lattices. The facile H₂O incorporation promoted by the presence of Ba is the necessary step to ensure a high proton conductivity as suggested by the water incorporation reaction:

$$H_2O + V_o^{\bullet\bullet} + O_O^\times \leftrightarrow 2OH_O^\bullet \qquad (1)$$

The proton conductivity of Ba- and Mg-doped LaGaO$_3$ (LBGM) has not been reported previously. In this study, we report the proton conductivity and water incorporation behavior of LBGM as a function of a wide range of water vapor pressure for the first time.

EXPERIMENTAL

Sample preparation

By a solid-state reaction method, the starting chemicals of La$_2$O$_3$ (>99.99% purity, Alfa Aesar), BaCO$_3$ (99.8% purity, Alfa Aesar), Ga$_2$O$_3$ (>99.99% purity, GFI) and MgO (>99.99% purity, Alfa Aesar) were weighed according to the stoichiometry of La$_{1-x}$Ba$_x$Ga$_{0.80}$Mg$_{0.20}$O$_{3-\delta}$ (x varies from 0.05 to 0.20).

To ensure the accuracy of the stoichiometry of La in powders, La$_2$O$_3$ was pre-calcined at 1000°C for 5 hours prior to actual weighing to remove any non-oxide components. After pre-calcination, the La$_2$O$_3$ powder was weighed shortly after it was taken out of furnace at 600°C. The weighed powders were mixed in an agate mortar with the aid of acetone. Pellets were subsequently pressed under a pressure of 200 MPa and initially fired at 1250°C for 10 hours. The partially reacted samples were then broken up, reground and ball-milled before cylindrical bars (ϕ5 mm in diameter and 10–12 mm in length) were pressed and finally sintered at 1450°C for 5 hours.

Characterization

The FEI Quanta 200 Environmental Scanning Electron Microscope (ESEM) was used to reveal microstructures of the samples. The phase purity of the final products were examined by powder X-ray diffraction (PXRD) using an X-ray diffractometer (D/max-A, Rigaku, Japan) equipped with graphite-monochromatized CuKα radiation (λ=1.5418 Å) at a scanning rate of 5° min^{-1} in a 2θ range from 20° to 80°. The patterns were analyzed with JADE (MDI) software.

The conductivity of the as-sintered bar samples was measured by AC impedance spectroscopy technique using a Solartron 1260 Frequency Response Analyzer and 1287 Electrochemical Interface. The frequency swept during impedance measurements varied from 0.1Hz to 1.2 MHz and accompanied by an AC perturbation amplitude of 10 mV. The impedance cell was configured in a symmetric fashion using two identical Ag electrodes. The isothermal proton conductivity study at different water vapor partial pressures was carried out at 600°C with either 50 sccm air or with 50 sccm H$_2$ as the carrier gas. The proton conductivity was also studied as a function of temperature at a fixed water partial pressure.

RESULTS AND DISCUSSION

Microstructural examination

The surface SEM micrographs of LBGM series (from x=0.05 to 0.20) are shown in Fig.1. La$_{0.95}$Ba$_{0.05}$Ga$_{0.80}$Mg$_{0.20}$O$_{3-\delta}$ and La$_{0.90}$Ba$_{0.10}$Ga$_{0.80}$Mg$_{0.20}$O$_{3-\delta}$ have a dense structure with well sintered grains while La$_{0.85}$Ba$_{0.15}$Ga$_{0.80}$Mg$_{0.20}$O$_{3-\delta}$ and La$_{0.80}$Ba$_{0.20}$Ga$_{0.80}$Mg$_{0.20}$O$_{3-\delta}$ are not fully dense. As the barium content increases, vaporization/condensation occurred during the sintering process becomes more severe, causing poorly sintered grains. In this study, x=0.10 is considered to be the upper doping limit for La$_{1-x}$Ba$_x$Ga$_{0.80}$Mg$_{0.20}$O$_{3-\delta}$ series.

Fig.1. Surface morphologies of La$_{1-x}$Ba$_x$Ga$_{0.80}$Mg$_{0.20}$O$_{3-\delta}$
(a) x=0.05 (b) x=0.10 (c) x=0.15 and (d) x=0.20

XRD Examination

The XRD patterns for La$_{1-x}$Ba$_x$Ga$_{0.80}$Mg$_{0.20}$O$_{3-\delta}$ are shown in Fig. 2. Expect that La$_{0.90}$Ba$_{0.10}$Ga$_{0.80}$Mg$_{0.20}$O$_{3-\delta}$ is single-phase perovskite, all other samples contain impurities. When x=0.05, the impurity is LaBa$_2$Ga$_{11}$O$_{20}$ with an unknown phase that cannot be indexed; when x≥0.15, the impure phases are LaBaGa$_3$O$_7$ and LaBaGaO$_4$, similar to those impurities LaSrGa$_3$O$_7$ and LaSrGaO$_4$ in LSGM [5].

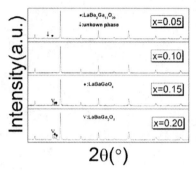

Fig.2. XRD of La$_{1-x}$Ba$_x$Ga$_{0.80}$Mg$_{0.20}$O$_{3-\delta}$

Electrical properties

Fig. 3 shows the bulk conductivity data for La$_{1-x}$Ba$_x$Ga$_{0.80}$Mg$_{0.20}$O$_{3-\delta}$ in air containing water partial pressure of 0.065atm. As x increases, the bulk conductivity first increases and peaks at x=0.10 and then decreases. Purer phase and denser microstructure at x=0.10 are attributed to its highest conductivity. Table 1 lists the active energy La$_{1-x}$Ba$_x$Ga$_{0.80}$Mg$_{0.20}$O$_{3-\delta}$ in this condition. Active energy increases with an increasing x. Overall, the averaged activation energy of La$_{1-x}$Ba$_x$Ga$_{0.80}$Mg$_{0.20}$O$_{3-\delta}$ series is typical for oxide-ion conduction, inferring that La$_{1-x}$Ba$_x$Ga$_{0.80}$Mg$_{0.20}$O$_{3-\delta}$ grain is more oxide-ion conducting.

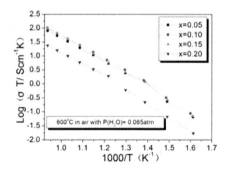

Fig.3. Arrhenius plot of bulk conductivity for La$_{1-x}$Ba$_x$Ga$_{0.80}$Mg$_{0.20}$O$_{3-\delta}$ in wet air

Table 1 Average activation energy of La$_{1-x}$Ba$_x$Ga$_{0.80}$Mg$_{0.20}$O$_{3-\delta}$ in wet air

x	0.05	0.10	0.15	0.20
E$_a$ (eV)	0.903	0.908	0.921	0.928

According to eq. (1), it is obvious that the availability of oxygen vacancies in proton conducting oxides is critical to achieving a high protonic conductivity. The $V_O^{\bullet\bullet}$ can be created by sub-valent acceptor doping on cation sites. In this study, B site doping is fixed at 20%, so the A site doping amount is important. The defect reaction on A site can be expressed as following:

$$2BaO \xrightarrow{La_2O_3} 2Ba'_{La} + 3O_O^x + V_O^{\bullet\bullet} \qquad (2)$$

As the doping content (x) increases, more oxygen vacancies would be created, which can react with water molecules according to equation (1) to produce proton species $OH_O^{\bullet}$. Further increasing the doping content would not necessarily lead to a better electrical performance because of the appearance of impurity phases LaBaGa$_3$O$_7$ and LaBaGaO$_4$ segregated to block ionic migration.

From Fig.1 to Fig.3, it is evident that La$_{0.90}$Ba$_{0.10}$Ga$_{0.80}$Mg$_{0.20}$O$_{3-\delta}$ is the best composition among the series. Therefore, the following characterizations have been mainly focused on La$_{0.90}$Ba$_{0.10}$Ga$_{0.80}$Mg$_{0.20}$O$_{3-\delta}$. The evolution of AC impedance spectra as a function of water vapor partial pressure with air and H$_2$ as the balancing gases is shown in Fig. 4. The overall spectra generally consist of contributions from inductance, grain, grain-boundary and electrode process. The highest-frequency inductance effect with positive imaginary component is an indicator of interferences from the measuring leads subject to a magnetic field caused by the furnace (a coiled heater), while the lowest-frequency spectrum is related to the Ag/LBGM electrode interface. At the intermediate frequency, the semicircle is related to grain-boundary contribution.

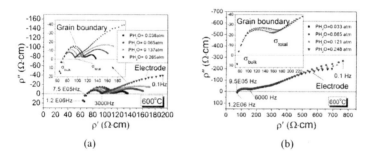

(a)   (b)

Fig.4. Impedance spectra of La$_{0.90}$Ba$_{0.10}$Ga$_{0.80}$Mg$_{0.20}$O$_{3-\delta}$ measured at 600°C as a function of water vapor partial pressure with (a) air and (b) H$_2$ as the balancing gases

Extracted from Fig. 4, the conductivities of bulk and grain-boundary are plotted as a function of water vapor partial pressure with both air and H$_2$ as the balancing gases in Fig. 5. In air-H$_2$O atmosphere, the grain-boundary conductivity shows not only persistently higher value than the bulk, but also a noticeable increase with water partial pressure; the bulk conductivity instead remains relatively independent on water vapor partial pressure. This finding suggests that the grain-boundary composition be proton conducting in oxidizing atmospheres. In H$_2$-H$_2$O atmosphere, neither bulk nor grain-boundary conductivity depends strongly with water vapor partial pressure, indicating the grain-boundary phase becomes non-proton conducting in reducing atmospheres. Furthermore, grain-boundary conductivity in reducing atmosphere is lower than that of bulk.

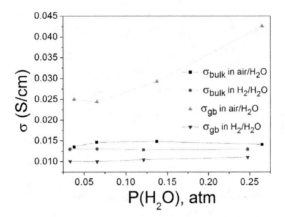

Fig.5. Bulk and grain boundary conductivity for La$_{0.90}$Ba$_{0.10}$Ga$_{0.80}$Mg$_{0.2}$O$_{3-\delta}$ in wet air and wet H$_2$

The fact that higher conductivity in oxidizing atmospheres than in reducing atmospheres implies that p-type electron holes are involved in the charge transport via the following defect reaction:

$$\frac{1}{2}O_2 + V_O^{\bullet\bullet} = O_o^x + 2h^\bullet \qquad (3)$$

CONCLUSIONS

The proton conductivity of a perovskite-structured La$_{1-x}$Ba$_x$Ga$_{0.80}$Mg$_{0.20}$O$_{3-\delta}$ (x varies from 0.05 to 0.20) were studied. When x=0.10, the sample is a dense ceramic with single phase that also exhibits the highest ionic conductivity. The proton conduction is noticeably found along the grain-boundary in oxidizing atmospheres whereas no such a grain-boundary effect is observed in reducing atmosphere. In either case, the bulk conductivity shows less dependence on water vapor partial pressure or proton conduction.

REFERENCES

[1] K. D. Kreuer, Proton Conductivity: Materials and Applications, *Chem. Mater*, **8**, 610-641(1996).
[2] A. Grimaud, J. M. Bassat, F. Mauvy, P. Simon, A. Canizares, B. Rousseau, M. Marrony, J. C. Grenier. Transport properties and in-situ Raman spectroscopy study of BaCe$_{0.9}$Y$_{0.1}$O$_{3-\delta}$ as a function of water partial pressures, *Solid State Ionics*, **191**, 24-31(2011).
[3] X. Li, N. Xu, L. Zhang, K. Huang, Combining proton conductor BaZr$_{0.8}$Y$_{0.2}$O$_{3-\delta}$ with carbonate: Promoted densification and enhanced proton conductivity, *Electrochemistry Communications*, **13**, 694-697(2011).
[4] T.Ishihara, H. Matsuda, Y. Takita, Doped LaGaO$_3$ Perovskite Type Oxide as a New Oxide Ionic Conductor, J. Am. Chem. Soc. **116**, 3801-03 (1994).

[5] K. Huang, R. Tichy, J. B. Goodenough, Superior Perovskite Oxide-Ion Conductor; Strontium- and Magnesium-Doped LaGaO$_3$:I, Phase Relationships and Electrical Properties, *J. Am. Ceram. Soc.* **81**, 2565-75 (1998).

[6] G. Ma, F. Zhang, J. Zhu, G. Meng, Proton Conduction in La$_{0.9}$Sr$_{0.1}$Ga$_{0.8}$Mg$_{0.2}$O$_{3-\alpha}$, *Chem. Mater,* **18**,6006-11 (2006).

[7] S. Li, F. Schonberger, P. Slater, La$_{1-x}$Ba$_{1+x}$GaO$_{4-x/2}$: a novel high temperature proton conductor, *Chem. Commun.*, 2694-95 (2003).

[8] F. Giannici, D. Messana, A. Longo, A. Martorana, Crystal Structure and Local Dynamics in Tetrahedral Proton-Conducting La$_{1-x}$Ba$_{1+x}$GaO$_4$, *J. Phys. Chem. C*, **115**, 298-304(2011).

SILVER-PALLADIUM ALLOY ELECTRODES FOR LOW TEMPERATURE SOLID OXIDE ELECTROLYSIS CELLS (SOEC)

Michael Keane, Prabhakar Singh
Center for Clean Energy Engineering, University of Connecticut
Storrs, CT, USA

INTRODUCTION

Large quantities of hydrogen i required for the upgrade and reforming of petroleum products. Potential large-scale hydrogen applications also include the production of synthetic hydrocarbon fuels via the Fischer-Tropsch process, and its direct use as a transportation fuel in emerging hydrogen fuel cell vehicles. Currently, 96% of the commercial hydrogen is produced by the conventional steam reforming and the partial oxidation of abundant natural gas and liquid hydrocarbons [1, 2]. Since both of the above processes produce significant carbon foot print and emissions, development of large-scale hydrogen production technologies with reduced emissions of greenhouse gases is of great interest to a wide variety of industries. Hydrogen produced from water electrolysis, as opposed to the reformation of hydrocarbons, offers the potential to be carbon-free when paired with a renewable energy source such as wind. The high temperature electrolysis of steam using solid oxide electrolysis cells (SOEC) have also been reported to show thermodynamic efficiency in excess of 50% [3].

An SOEC, schematically shown in Figure 1, comprises of a dense oxygen ion conducting electrolyte layer and two porous electrodes. This mode of operation, whereby energy is used to convert water into hydrogen and oxygen, is essentially the reverse of the electrode processes operating in a solid oxide fuel cell (SOFC). In both SOECs and SOFCs, the electrolyte is typically a dense yttria-stabilized zirconia (YSZ), and the fuel electrode is usually a porous nickel-YSZ cermet composite. The air electrode is commonly a porous lanthanum strontium manganite (LSM) or a LSM-YSZ composite [4]. For larger scale hydrogen production, the cells are series connected, arranged in a stack, and separated by electronically conducting interconnects [5].

One of the most important developmental barriers in SOEC technology is that of long-term degradation. Cell degradation, responsible for as high as 40% reduction in hydrogen production rate in 2000 hour tests with SOEC stacks has been reported [7] and a number of causes for the observed degradation have been identified [5]. The degradation has been classified into two broad groups: (a) electrode poisoning due to gas phase contamination, and (b) gradual delamination and separation of the electrode at the anode-electrolyte interface. The performance degradation associated with chromium poisoning of the anode [5] and silica poisoning of the cathode [8] are well documented for high temperature electrochemical systems including SOEC and SOFC. It has been suggested that the transport of chromium from steel interconnects to the electrode-electrolyte interface could lead to the air electrode delamination [17]. The observations remain inconclusive as tests conducted in the absence of chromium-containing materials have also resulted in the electrode delamination [10]. Anode delamination from the electrolyte results in a current constriction and increased ohmic loss [5, 9–12] and several processes for the delamination of the electrode ranging from high oxygen pressure development, electrolyte grain boundary separation, and morphological changes in the LSM anode have been postulated. The anode delamination and interface separation has been considered as the largest contributor to cell performance degradation [5].

Lowering the operating temperature of solid oxide cells is an effective method of reducing the materials interactions that lead to performance degradation, thus improving long-term performance. Lower temperatures have several other benefits including shorter heat-up time, reduced stack heating requirements, and lower fabrication costs. Materials for interconnects and balance-of-plant components have less stringent requirements, further reducing the cost. However, many common solid oxide cell materials have poor performance at lower temperatures, such as the poor reaction kinetics of LSM and low ionic conductivity of YSZ [1].

Silver-based electrodes have been investigated for use in lower temperature solid oxide cells (550-750 °C) due to high catalytic performance even at lower temperature, in addition to good electronic conductivity and oxygen permeability [2]. While LSM and similar ceramic electrodes are known to form insulating zirconate products at the electrode-YSZ electrolyte interfaces, there are no known reaction products between silver and YSZ. However, silver is typically used as a component of silver-ceramic composite electrodes to avoid the oversintering and evaporation during sintering of pure silver [3]. Stable performance for $Ag-Er_{0.4}Bi_{1.6}O_3$-YSZ composite electrode has been reported at 650 °C [4]. Promising results have also been reported for silver-doped or infiltrated electronic conductors such as LSM [5], as well as ionic conductors such as SDC [6].

In order to evaluate the potential of silver-based electrodes in SOEC, the electrochemical performance and stability of symmetric cells with silver-palladium electrodes have been investigated in this study. Unlike the conventional full-cell configuration, the selected test configuration simplifies the cell assembly, eliminates sealing requirements as both electrodes of the cell are exposed only to air. The electrochemical reactions associated with the oxygen reduction at the cathode remains identical to the cathodic reaction in an SOFC, while the oxygen ion oxidation at the anode is identical to the anodic reaction in an SOEC. The cell arrangement also provides a simple configuration where air electrode behavior in both SOFC and SOEC conditions can be compared. The use of symmetric cells also eliminates the possibility of degradation at the SOEC cathode due to gas phase nickel transport, coarsening and contamination resulting from silica. The symmetric cells were electrochemically tested at various operating voltages to measure the overall electrochemical degradation and examine chemical and structural changes at the electrode–electrolyte interfaces.

EXPERIMENTAL

25 mm diameter symmetric button cells (Air/Ag-Pd//YSZ//Ag-Pd/Air) consisting of 190 μm thick $(ZrO_2)_{0.92}(Y_2O_3)_{0.08}$ (YSZ) electrolyte (Fuel Cell Materials) and 10 μm thick silver-palladium electrode (ESL Electroscience, 15% Pd) were fabricated for electrochemical testing. Concentric electrodes of 10 mm diameter were applied on both sides of the electrolyte disc by screen printing technique using 105 mesh screen. Electrodes were subsequently bisque-fired in air for 1 hour at 450 °C to volatilize organics, and then sintered for 1 hour at 850 °C to firmly adhere the electrode to the YSZ substrate. The electrochemical active area of the cell electrodes was calculated to be 0.8 cm^2. Silver screen current collector (Alfa Aesar, 50 mesh) with silver wires (Alfa Aesar, 0.20 mm) were attached to each electrode using a second layer of silver-palladium paste. Sintering of the current collector was performed in air for 1 hour at 850 °C. Schematic of the experimental test set up is shown in Figure 1. A 2.5 cm diameter alumina tube was used to support the symmetric cell assembly and the leads from a multi-channel potentiostat (VMP2, Bio-Logic) were attached to the assembled button cell. A type K (Chromel-alumel) thermocouple was placed within 5 mm of the cell to monitor the operating temperature.

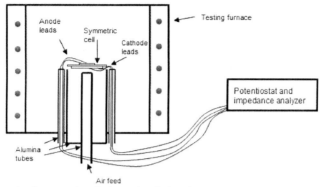

Figure 1 Schematic of symmetric cell electrochemical testing apparatus.

The cells were heated to 750 °C at 3 °C/min in flowing air. Using the potentiostat, a predetermined constant voltage was applied for 100 hours and the cell current was recorded every 60 seconds through the duration of the test. Impedance measurement was performed at four hour intervals using a 10 mV alternating current in the frequency range from 100 mHz to 200 kHz. Experiments were repeated several times at each imposed voltage conditions to ensure repeatability. All tests were performed under flowing air conditions with flow rate maintained at 300 sccm. Although typical SOECs are operated at 0.3 to 0.4 volts above OCV [6], cells in the current study were tested in a wide voltage range (0 to 0.8 volts) to assess the degradation under simulated nominal and accelerated cell operating conditions.

Tested cells were subsequently analyzed for structural, morphological and chemical changes. Bulk electrode and electrode–electrolyte interfaces were examined for morphological changes and reaction products formation. Interfaces were further examined after dissolution of the electrode in dilute nitric acid at room temperature for two hours. An FEI Quanta 250 FEG scanning electron microscope (SEM) and EDAX (attached to ESEM) energy-dispersive X-ray spectroscopy (EDS) was used for the morphological and elemental distribution study. A Bruker AXS D-8 Advance X-ray diffractometer (XRD) was used for the identification of compounds present in both pre and post tested cells.

RESULTS AND DISCUSSION

After screen-printing and drying the silver-palladium paste on a YSZ substrate, the morphology in Figure 2a is observed. The average particle size is approximately 200 nm. After bisque-firing, a bimodal particle size distribution is observed as in Figure 2b. Average particle sizes are 200 nm and 1.5 μm. After sintering, average particle size increases to 7 μm (Figure 3c). Porosity is estimated at a constant 5% for all three conditions, showing that the primary effect of high-temperature treatment is particle sintering rather than densification. Final sintered thickness is about 10 μm.

a)   b)   c)

Figure 2 SEM images of screen-printed, untested silver-palladium electrode at 30000 times magnification. a) Dried at 80 °C for 1 hour. b) Bisque-fired at 450 °C for 1 hour. c) Sintered at 850 °C for 1 hour.

Figure 3 shows a typical impedance spectra exhibited by the symmetrical cell at 0.8 volts. The leftmost x-intercept represents cell ohmic resistance and the diameter of the semi-circle represents the non-ohmic resistance. Ohmic resistance is primarily contributed by the YSZ electrolyte because the cells are electrolyte-supported. Non-ohmic resistance includes oxygen reduction and re-oxidation reactions at the electrode/electrolyte interface, oxygen absorption, surface and bulk diffusion [18, 19].

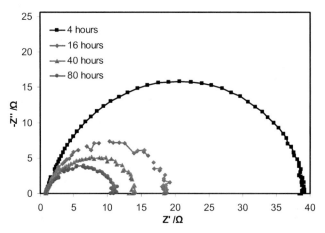

Figure 3 Nyquist plot of impedance spectra from a symmetric cell tested at 0.8 volts. For clarity, only 4 spectra are shown.

The ohmic and non-ohmic resistances taken from the Nyquist plot are plotted as a function of time in Figures 4 and 5, respectively. For comparison, results from cells tested with pure LSM electrodes are also plotted [20]. Ohmic resistance of cells with silver-palladium is initially three times higher (0.9 Ω) than that of cells with LSM (0.3 Ω) (Figure 4). A factor of two difference can be explained by the lower testing temperature of cells with Ag-Pd causing a decrease in YSZ conductivity. The remaining 0.3 Ω discrepancy is therefore attributed to ohmic resistance of the Ag-Pd electrodes. The ohmic resistance of cells with Ag-Pd is stable over the testing period, even showing a slight decrease with time. In contrast, the cells with LSM show significant degradation over this time period, even surpassing the resistance of the Ag-Pd cell after 80 hours. Therefore, Ag-Pd electrodes have significantly improved stability compared to LSM, from the standpoint of ohmic resistance.

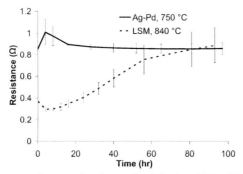

Figure 4 Ohmic resistances of symmetric cells tested at 0.8 volts with Ag-Pd and LSM electrodes as functions of time.

Non-ohmic resistance of cells with Ag-Pd electrodes begins very high but rapidly decreases with time, reaching a minimum after about 40 hours (Figure 5). A similar but less significant decrease is seen in cells with LSM electrodes. This so-called activation effect is related to passivation of the electrode surface which is removed during electrochemical testing [21]. Even after activation, non-ohmic resistance of the cells with Ag-Pd is more than an order of magnitude higher than that of cells with LSM. Non-ohmic resistances of LSM cells begin to increase after the activation period. These increases are evidence of degradation related to the diffusion, absorption, and oxidation/reduction of oxygen at the electrodes. The behavior of non-ohmic resistances of the Ag-Pd cells showed considerable variation; some of the cells showed slight degradation after the activation period, while others showed continual slight activation for the duration of the test.

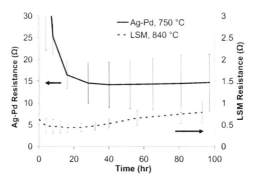

Figure 5 Non-ohmic resistances of symmetric cells tested at 0.8 volts with Ag-Pd and LSM electrodes as functions of time.

After electrochemical testing, the Ag-Pd-containing cells showed no changes upon initial visual inspection, and the electrodes were still firmly adhered to the YSZ electrolyte. SEM observation was performed on the electrodes for observation of morphological changes. While the cathodes retain their porous structure in all cases, the anodes of some of the cells show considerable densification (Figure 6). The anodes of cells that have non-ohmic degradation tended to densify, while those of cells that do not have such degradation tended to retain their porous structure. Ag-Pd densification should lead to a reduction in triple phase boundary (TPB) length, which reduces the sites for oxygen ion reoxidation and increases the observed non-ohmic resistance.

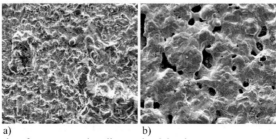

a)   b)

Figure 6 Ag-Pd anodes of two symmetric cells tested at 0.8 volts.

In order to observe the Ag-Pd/YSZ interface, the Ag-Pd was dissolved in dilute nitric acid. The inactive YSZ surface (Figure 7a has few features except for the grain boundaries and some sparsely distributed small pores. The cathode and anode-side active YSZ surfaces are shown in Figures 7b and 7c respectively. These surfaces were in contact with the Ag-Pd electrodes during electrochemical testing. These surfaces have minimal morphological changes compared with the inactive YSZ surface. However, there are a few remaining particles and evidence of a slight imprint from the electrode in many areas. EDS analysis shows some remaining palladium, which is less soluble than silver in nitric acid.

On the anode side, there are a few isolated areas concentrated at the edges of the active area that showed severe morphological changes (Figure 7d). The YSZ grain boundaries have become separated, with the separation extending 1-2 μm from the active area. The YSZ grains have also developed a complex network of interconnected porosity. Because this surface damage only occurs in a few small areas at the edge of the active area, it is hypothesized that current constriction leading to high current density may play a role.

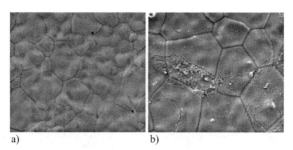

a)   b)

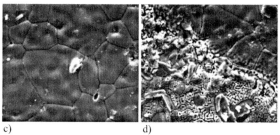

c)   d)

Figure 7 YSZ surface morphologies of symmetric cells after applying 0.8 volts for 100 hours at 750 °C and dissolving Ag-Pd electrodes in nitric acid (30000 times magnification). a) Inactive surface. b) Active surface on cathode side. c) Active surface on anode side. d) Edge of active surface on anode side.

X-ray diffraction was also performed on the active surfaces. Both the anode and cathode sides showed pure YSZ with no evidence of other phases (Figure 8). The absence of interfacial reaction products is in contrast with perovskite electrodes such as LSM. Electrically insulating lanthanum zirconate is known to form at LSM/YSZ interfaces in solid oxide cells [22]. The lack of insulating phases forming at the Ag-Pd/YSZ interface could explain the stability of the ohmic resistance seen in Figure 4.

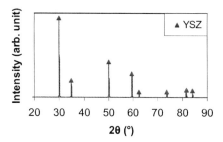

Figure 8 X-ray diffraction pattern of anode-side YSZ surface of symmetric cell after applying 0.8 volts for 100 hours at 750 °C and dissolving Ag-Pd electrodes in nitric acid.

In order to account for the high ohmic resistance of the cells as seen in Figure 4, the expected resistance based on oxygen atom flux through a dense silver film was calculated based on data from Fromm and Gebhardt [23] (Figure 9). The resistance of a 10 mm circle of 10 μm thick dense silver at 750 °C is about 1 Ω. For two identical dense electrodes, the ohmic contribution would be 2 Ω. Since the electrodes are not completely dense, the true contribution should be less than 2 Ω. This is because some oxygen will diffuse through the porous as a gaseous species rather than through the silver bulk as an atomic species. Thus, the limitation of atomic oxygen flux through the dense portions of the electrodes can explain the 0.3 Ω discrepancy in ohmic resistance in Figure 4.

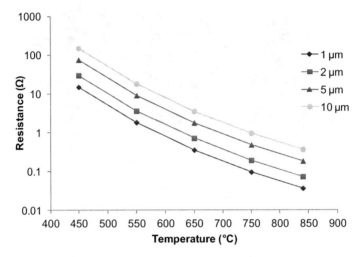

Figure 9 Calculated maximum oxygen flux in terms of current density through a dense silver layer.

The porosity and surface damage in the YSZ electrolyte (Figure 7d) on the anode side may be explainable through the buildup of extremely high oxygen pressures at the anode-electrolyte interface. Potential interfacial pressures were calculated based on equations by Virkar et al. [24] (Figure 10). The calculations are based on a dense silver film. The electrode/electrolyte thickness ratio is about 0.05 in the tested cells, so oxygen pressures several orders of magnitude are obtainable in regions where the electrode is dense. The nucleation of high-pressure oxygen gas in the electrolyte near the anode-electrolyte interface could thus be responsible for the observed morphological damage in the YSZ. Since the damaged regions are only observed near the edge of the active area, it is possible that the current becomes concentrated in these regions due to slight misalignment of the anode and cathode.

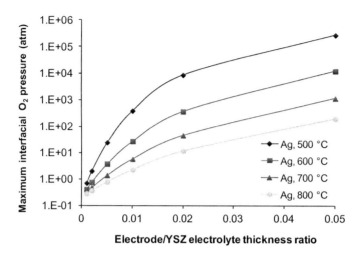

Figure 10 Calculated maximum oxygen pressure at anode-electrolyte interface during application of 0.3 volt bias.

CONCLUSION

Symmetric solid oxide cells with Ag-Pd electrodes and YSZ electrodes show relatively lower but predominantly stable performance when electrically tested with a 0.8 volt bias at 750 °C for 100 hours. The electrode-electrolyte interfaces are stable except for localized regions on the anode side which show YSZ surface damage. Performance stability is attributed to the morphological and chemical stability of the electrode-electrolyte interfaces. Lower performance is attributed to very low electrode porosity, which limits the oxygen flow and thus current. Low porosity is also responsible for high oxygen pressure in some areas of the anode-electrolyte interface, leading to morphological damage in the YSZ surface.

ACKNOWLEDGEMENT

The authors acknowledge financial support from ConocoPhillips and Idaho National Laboratory. Also acknowledged are Professor Steven Suib and his research group members Cecil King'ondu, Yongtao Meng, and Altug Poyraz from the UConn Department of Chemistry, as well as Professor Singh's research group members Dr. Manoj Mahapatra, Dr. Kailash Patil, Gavin Ge, Na Li, Sapna Gupta, and Keling Zhang.

REFERENCES
[1] R. Ramachandran and R.K. Menon. An Overview of Industrial Uses of Hydrogen. *International Journal of Hydrogen Energy* **23** (7) 593–598 (1999).
[2] R. Elder and R.W.K. Allen. *Progress in Nuclear Energy* **51** 500–525 (2009).
[3] J.E. O'Brien, M.G. McKellar, E.A. Harvego, and C.M. Stoots. High-temperature electrolysis for large-scale hydrogen and syngas production from nuclear energy – summary of system simulation and economics analyses. *International Journal of Hydrogen Energy* **35**, 4808–4819 (2010).

[4] C. Yang, A. Coffin, and F. Chen. High temperature solid oxide electrolysis cell employing porous structured $(La_{0.75}Sr_{0.25})_{0.95}MnO_3$ with enhanced oxygen electrode performance. *International Journal of Hydrogen Energy* **35**, 3221–3226 (2010).

[5] J.R. Mawdsley, J.D. Carter, A.J. Kropf, B. Yildiz, and V.A. Maroni. Post-test evaluation of oxygen electrodes from solid oxide electrolysis stacks. *International Journal of Hydrogen Energy* **34**, 4198–4207 (2009).

[6] J.E. O'Brien, C.M Stoots, J.S. Herring, and J.J. Hartvigsen. Performance of planar high-temperature electrolysis stacks for hydrogen production from nuclear energy. *Nuclear Technology* **158**, 118–131 (2007).

[7] V.I. Sharma and B. Yildiz. Degradation Mechanism in La0.8Sr0.2CoO3 as Contact Layer on the Solid Oxide Electrolysis Cell Anode. *Journal of the Electrochemical Society* **157** (3), B441–B448 (2010).

[8] S.D. Ebbesen, C. Graves, A . Hauch, S.H. Jensen, M. Mogensen. Poisoning of Solid Oxide Electrolysis Cells by Impurities. *Journal of the Electrochemical Society* **157** (10) B1419–B1429 (2010).

[9] A. Kaiser, E. Monreal, A. Koch, and D. Stolten. Reactions at the Interface $La_{0.5}Ca_{0.5}MnO_3$-YSZ/Al_2O_3 under Anodic Current. *Ionics* **2**, 184–189 (1996).

[10] A. Momma, T. Kato, Y. Kaga, and S. Nagata. Polarization Behavior of High Temperature Solid Oxide Electrolysis Cells (SOEC). *Journal of the Ceramic Society of Japan* **105** [5], 369–373 (1997).

[11] J. Guan, N. Minh, B. Ramamurthi, J. Ruud, J. Hong, P. Riley, and D. Weng. High Performance Flexible Reversible Solid Oxide Fuel Cell. GE Global Research Center, Torrance, CA., 2004-2006.

[12] H. Lim and A.V. Virkar. A study of solid oxide fuel cell stack failure by inducing abnormal behavior in a single cell test. *Journal of Power Sources* **185**, 790–800 (2008).

[17] R. Hino, K. Haga, H. Aita, K. Sekita. R&D on hydrogen production by high-temperature electrolysis of steam. *Nuclear Engineering and Design* **233** 363–375 (2004).

[18] T. P. Holme, R. Pornprasertsuk, F. B. Prinz. Interpretation of Low Temperature Solid Oxide Fuel Cell Electrochemical Impedance Spectra. *Journal of the Electrochemical Society* **157** (1) B64–B70 (2010).

[19] S.B. Adler. Limitations of charge-transfer models for mixed-conducting oxygen electrodes. *Solid State Ionics* **135** 603–612 (2000).

[20] M. Keane, A. Verma, P. Singh. LSM-YSZ Interactions and Anode Delamination in Solid Oxide Electrolysis Cells. Unpublished manuscript. (2012).

[21] S.P. Jiang, J.G. Love, J.P. Zhang, M. Huong, Y. Ramprakash, A.E. Hughes, S.P.S. Badwal. The electrochemical performance of LSM/ zirconia–yttria interface as a function of a-site non-stoichiometry and cathodic current treatment. *Solid State Ionics* **121** 1–10 (1999).

[22] A. Mitterdorfer and L.J. Gauckler. $La_2Zr_2O_7$ formation and oxygen reduction kinetics of the $La_{0.85}Sr_{0.15}Mn_yO_3$, $O_2(g)$|YSZ system. *Solid State Ionics* **111**, 185–218 (1998).

[23] E. Fromm, E. Gebhardt. *Gase und Kohlenstoff in Metallen.* Springer-Verlag: Berlin/Heidelberg 1976; p 679.

[24] A.V. Virkar, J. Nachlas, A.V. Joshi, and J. Diamond. Internal Precipitation of Molecular Oxygen and Electromechanical Failure of Zirconia Solid Electrolytes. *Journal of the American Ceramic Society* **73** [11], 3382–3390 (1990).

[1a] E.D. Wachsman, K.T. Lee. Lowering the Temperature of Solid Oxide Fuel Cells. *Science* **334** 935–939 (2011).

[2a] K. Sasaki, M. Muranaka, A. Suzuki, T. Terai. Determination of Oxygen Pathway in Silver Cathodes by Secondary-Ion Mass Spectroscopy Using Oxygen Isotope Exchange. *Journal of the Electrochemical Society* **158** (11) B1380–B1383 (2011).

[3a] Z. Wu, M. Liu. Ag-$Bi_{1.5}Y_{0.5}O_3$ Composite Cathode Materials for $BaCe_{0.8}Gd_{0.2}O_3$-Based Solid Oxide Fuel Cells. *Journal of the American Ceramic Society* **81** (5) 1215–1220 (1998).

[4a] M. Camaratta, E. Wachsman. Silver-bismuth oxide cathodes for IT-SOFCs Part II – Improving stability through microstructural control. *Solid State Ionics* **178** 1411–1418 (2007).

[5a] W. Zhou, Z. Shao, F. Liang, Z.-G. Chen, Z. Zhu, W. Jin, N. Xu. A new cathode for solid oxide fuel cells capable of *in situ* electrochemical regeneration. *Journal of Materials Chemistry* **21** 15343–15351 (2011).

[6a] Y. Lin, C. Su, C. Huang, J.S. Kim, C. Kwak, Z. Shao. A new symmetric solid oxide fuel cell with a samaria-doped ceria framework and a silver-infiltrated electrocatalyst. *Journal of Power Sources* **197** 57–64 (2012).

DEVELOPMENT OF IMPROVED TUBULAR METAL-SUPPORTED SOLID OXIDE FUEL CELLS TOWARDS HIGH FUEL UTILIZATION STABILITY

L. Otaegui[1], L.M. Rodriguez-Martinez[1], M. A. Alvarez[1], F. Castro[2], I. Villarreal[1]
[1]Ikerlan, Parque Tecnológico de Alava, Juan de la Cierva 1
Miñano 01510, Álava, Spain
[2]CEIT, Paseo Manuel Lardizabal 15
San Sebastián, Guipúzcoa, Spain

ABSTRACT

Tubular metal-supported SOFC technology has successfully been developed over the past years with the aim at small combined (CHP) and portable systems. First generation of cells have been successfully tested up to 2000 h under current loading and more than 500 thermal cycles at low humidification conditions (3% H_2O/H_2). However, good resistance against oxidation due to high fuel utilization was not a success. A special effort has been devoted to determine the reason for the catastrophic degradation observed during operation at high fuel utilization conditions. Individual tests performed for metal support, diffusion barrier layers and anode structured samples under high humidification atmospheres (50% H_2O/H_2, 800°C) have demonstrated that modifications in the diffusion barrier layer improve significantly the resistance to oxidation of the metal support, achieving more than 500 hours with almost no degradation. Furthermore, a second generation of cells that can operate at higher fuel utilization conditions for more than 700 hours have been successfully demonstrated.

INTRODUCTION

Fuel cells are electrochemical devices that convert chemical energy in fuels directly into electrical energy, promising power generation with high efficiency and low environmental impact.[1] The majority of SOFC developments to date focus on electrolyte-supported, cathode-supported or anode-supported cells in which the mechanical support is a brittle ceramic or cermet. In contrast, the metal-supported cell design utilizes ceramic layers only as thick as necessary for electrochemical function. The mechanical support is made from inexpensive and robust porous metal and the electrochemically active layers are applied directly to the metal support. This metal-supported cell design provides significant cost, manufacturing, abuse tolerance and operational advantages that make it a very promising candidate for commercialisation.[2]

Most developers favour ferritic stainless steel for the metal support which typically contain between 10.5 - 26 wt% Cr to form a continuous chromia scale.[2] However, the main problem concerning this type of cells is the diffusion of iron, chromium and nickel between the ferritic FeCr steel and nickel containing anodes during the high temperature sintering processes required for complete densification of the electrolyte. Diffusion of nickel into the ferritic substrate may cause its austenitisation, which would result in a thermal expansion coefficient mismatch compared to the other cell components and a reduction in the Cr diffusion in the alloy, altering the selective oxidation of Cr and resulting in poorer oxidation resistance. On the other hand, the diffusion of iron and especially chromium into the anode may cause the formation of oxide scales on the nickel particles during cell operation and hence, degradation of cell performance.[3, 4]

Several options have been considered by developers to solve this problem and manufacture metal-supported SOFCs. On the one hand, some of them use low temperature deposition techniques such as vacuum plasma spray (VPS),[5, 6] the lower cost low pressure plasma spray (LPPS)[5, 6] or atmospheric plasma spray (APS),[5-7] pulsed laser deposition (PLD)[8] or electrophoretic deposition (EPD).[9] On the other hand, some others infiltrate the electrocatalitically active Ni particles after the sintering process at high temperature[10-13] or select electrolyte components that densify at lower temperatures.[9] Anyway, some of those deposition techniques can be combined with the addition of a diffusion barrier layer (DBL) which avoids Fe, Cr and Ni interdiffusion between the metal substrate and the anode.[5] Moreover, the addition of this DBL could also be effective for high temperature sintering fabrication routes.[14]

Two main groups of DBL are proposed in the literature: ceria based and perovskite type materials. The ones that show the most promising results are the perovskite-type Sr-doped Lanthanum-Chromites ($LaCrO_3$) and Lanthanum-Manganites ($LaMnO_3$). Szabo et al.[6] have demonstrated more than 2000 hours with less than 1%/kh degradation using a simulated reformate gas and 15 thermal and subsequent 20 redox cycles without a severe degradation of the cell. Ceria based materials (CeO_2 and $Ce_{0.8}Gd_{0.2}O_2$) combined with a Cu interlayer also show encouraging results of more than 1000 hours with no diffusion of Ni or any of the alloying elements of Crofer 22 APU. The components self-grown on the surface by pre-oxidation of the metal support (Cr_2O_3 and Cr_2MnO_4) were proposed as an easy way to avoid the interdiffusion of the elements but not successful results were obtained.[3, 4, 15]

The present work summarizes the evolution of tubular metal-supported SOFC technology developed at Ikerlan where cells are obtained by co-sintering the metal support, together with the DBL, the anode and the electrolyte in non-oxidizing atmosphere. When pushing the performance of the first generation of cells, it was encountered a severe damage under atmospheres with high water vapour concentration operating at 800 °C. The state of the art performance of these cells, the identification of their principal causes of degradation and the development of a second generation of improved cells are the main topics summarized in this work.

EXPERIMENTAL

Tubular metal supported solid oxide fuel cells of 5 - 10 cm length and active areas of 4 - 16 cm^2 have successfully been prepared with low cost industrial scaleable routes. The DBL, nickel/yttria stabilized zirconia (Ni-YSZ) anode and thin yttria stabilized zirconia (YSZ) electrolyte are deposited on the metal support by dip coating and co-fired in non-oxidising atmosphere at 1350 - 1370 °C. The lanthanum doped strontium ferrite/samarium doped ceria (LSF-SDC) composite cathode is subsequently dip coated and fired *in situ* at 950°C prior to electrochemical characterisation. Two different diffusion barrier layers (G1-type and G2-type, proprietary) that define the cell generation (first and second generation respectively), were used in this study.

Galvanostatic and thermal cycling durability tests under low fuel utilization (FU) conditions were performed in order to determine the stability of first generation cells. Those tests were carried out using inert alumina housing and metallic parts for galvanostatic and thermal cycling tests, respectively. Atmospheric air was used at the cathode side and excess humidified hydrogen gas (3% H_2O) at the anode side. Pt paste and mesh and Ni mesh were used respectively in the cathode and anode side for current collection and Ni wires were welded directly to the tubular cell and to the anode interconnect to check stability of both single cell and cell connection. IV and EIS measurements at 800 °C were carried out for monitorization every 200 hours galvanostatic test and 50 thermal cycles and basic experimental conditions have been based on the FCTestnet Procedures.[16] Due to modifications in furnace insulation during the thermal cycling durability test, thermal cycles were performed between 800 °C and a minimum variable temperature below 180 °C in the first 140 cycles and below 80 °C after 140th cycle. Heating and cooling rates were set at 10 °C/min.

Galvanostatic tests under higher fuel utilization conditions were performed using the same experimental setup mentioned above for galvanostatic steady state operation tests under low fuel utilization and the increase in fuel utilization conditions was achieved by lowering the excess 3% humidified hydrogen gas flow from 200 ml/min·cm^2 to 5.6 ml/min·cm^2. A cell with G2-type DBL that was stable for 650 hours in low gas flow conditions was subsequently exposed to simulated high fuel utilization conditions (hydrogen with 50% humidification). The humidified fuel atmosphere was produced by bubbling pure hydrogen in water at 82 °C (saturated steam at 0.5133 bar) using 2 bottle humidification system from Fuel Cell Technologies. The outflow was checked out beforehand measuring its relative humidity and temperature by means of Vaisala humidity and temperature transmitter HMT337.

As Fe, Cr and Ni interdiffusion that happens between the metal support and the anode was determined to be the cause of failure when cells were tested under low gas flow, this interaction was studied on metal support/Ni and metal support/DBL/anode multilayer structures. First, the effect on Fe, Cr and Ni interaction as a function of temperature and fuel humidification was analysed using metal support/Ni structures. Then, anode side components samples (composed of metal support, DBL and anode) were tested under simulated SOFC operating conditions (hydrogen with 50% humidification). The tubular samples (2 cm long) consisting of metal support/Ni and metal support/DBL/anode multilayer structures were prepared under identical conditions used for fabrication of complete cells. Sintering temperatures had been slightly adjusted in order to achieving similar porosities to those for complete cells. Ni paste and mesh were used for current collection and experiments were run at 800 °C, under constant current load (500 mA/cm^2).

The microstructure of the samples was examined in a FEI 200 Quanta FEG scanning electron microscope (SEM) and energy dispersive X-ray analysis (EDS).

RESULTS AND DISCUSSION

State of the Art of a First Generation of Tubular Metal-Supported SOFCs in Ikerlan

During the last years an especial effort has been done in the development of a first generation of robust tubular metal supported SOFC technology. More than 2000 hours under constant current load were achieved in low fuel utilization conditions when 3% humidified hydrogen in excess was fed as shown in Figure 1. The failure in cell performance shown in Figure 1 after 1500 hours is believed to be due to a defective sealing system (more details are reported elsewhere).[17, 18]

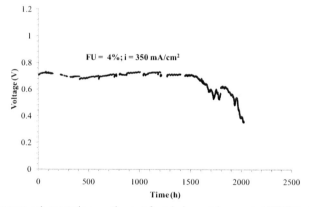

Figure 1. Long term galvanostatic operation test for tubular metal-supported SOFC tested at 800 °C. Fuel: 3% humidified pure H$_2$, FU = 4%.

After improvements in sealing materials and design, more than 500 thermal cycles were successfully performed as demonstrated in Figure 2, where the shadowed area corresponds to open circuit potential (OCP) data during cooling and heating. Due to high density of data-points, values at

800 °C and minimum temperature per cycle (black points on the top and the bottom of the graph) are highlighted. The graph shows how the open circuit characteristics at 800 °C remain almost constant after more than 500 thermal cycles and 2000 hours under cyclic operation.

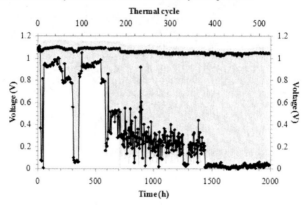

Figure 2. Evolution of OCP of a tubular metal-supported cell upon thermal cycling. Fuel: 3% humidified hydrogen, FU = 5%.

Having achieved good resistance in terms of durability and thermal cycling under low fuel utilization (FU = 4-5%), first cells were tested under higher fuel utilization conditions by progressively decreasing the fuel gas flow. Figure 3 shows the voltage evolution stability for one of those cells in comparison to the ones tested with excess hydrogen flow (the cell represented in Figure 1 was chosen as a comparison). During this experiment 3% humidified hydrogen flow and voltage were set constant at 5.6 ml/min·cm^2 and 0.8 V respectively, being the applied current 120 mA/cm^2 and equivalent to a small increase in fuel utilization from 4% up to 17%. After a short exposure time of 10 minutes to 17% fuel utilization, the voltage dropped dramatically so current load was then readjusted to 82 mA/cm^2 to maintain the initial operation conditions of 0.8 V, lowering fuel utilization form 17% to 12%. After 2 hours operating at 12% fuel utilization, current loading was again readjusted to 52 mA/cm^2, corresponding to 0.8 V and 9% fuel utilization, conditions that were then maintained for 10 hours until the complete failure of the cell.

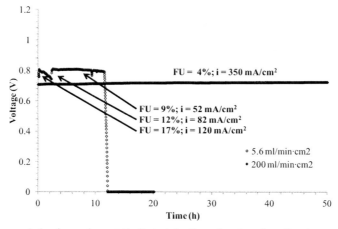

Figure 3. Voltage evolution for a galvanostatically tested cell as a function of gas flow (open symbols: cell tested under 5.6 ml/min·cm² and FU = 17%; 12% and 9%; closed symbols: cell tested under 200 ml/min·cm² excess gas flow and FU = 4%).

Post test SEM analysis of the cell cross section (Figure 4) shows a severe degradation near the electrochemically active area even at fuel utilization conditions as low as 9%. Most of the degradation is located at the interface between metal support and DBL. For this reason, independent studies were performed on metal support/Ni current collector and metal support/DBL/anode multilayer structures. This strategy was set to determine the component that limits the stability of the structure and identify materials and solutions for robust SOFC cells towards the objective of 70% fuel utilization.

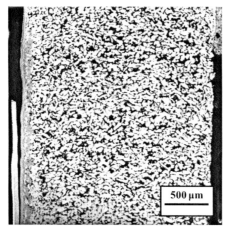

Figure 4. Cross section of the cell tested under low gas flow.

Analysis of the Interaction Between the Metal Support and the Diffusion Barrier Layer

EDS analysis on previously tested samples at low gas flow conditions indicates the presence of Fe, Cr and Ni interdiffusion between the ferritic substrate and the anode. Further, detailed experiments were performed in order to determine when that diffusion processes take place. For that purpose, several porous metallic substrates coated with Ni paste were fired in non-oxidising atmosphere at 1350 °C (sintering temperature) and 800 °C (operating temperature). After processing, samples were exposed to 3% and 50% humidified hydrogen atmospheres at 800 °C. Previous work had demonstrated that degradation is not catastrophic in the case of the bare metal support when exposed to this atmosphere,[18] so the effect of the Ni coating and its interaction with metal substrates under several conditions were studied. After exposure to the humidified atmosphere and operating temperature over time, the samples were observed by FEG-SEM.

Figure 5 a) and b) correspond to Ni coated FeCr samples sintered at 1350 °C before and after operation under 50% humidified atmosphere. Images correspond to the interface between Ni and the ferritic alloy. Ni that diffused through the FeCr substrate was detected at 700 μm far from the Ni source in the as sintered sample, while a 500 μm thick Fe and Cr oxide layer had grown in the FeCr and Ni contact area after 70 hours exposure to the 50% H_2O/50% H_2 atmosphere, demonstrating that catastrophic oxidation is caused by the Ni that diffused to the FeCr substrate during the sintering process. It is worth mentioning that the area where Ni is not coating the surface of the porous metal substrate remains free of the severe oxidation that occurs in the regions where direct contact exists.

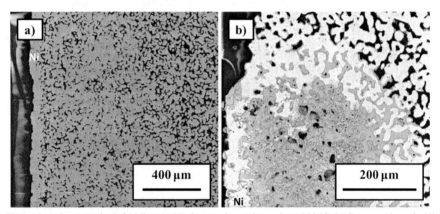

Figure 5. a) Cross section of a Ni coated FeCr sample co-sintered at 1350 °C. b) Cross section of a Ni coated FeCr sample co-sintered at 1350 °C after exposure during 70 h to 50% humidified hydrogen atmosphere at 800 °C.

In the case where Ni coated FeCr substrates were sintered at 800 °C and then exposed to operation temperature (800 °C), catastrophic degradation in the FeCr/Ni interface was only observed in the samples exposed to 50% humidified H_2 (Figure 6 b)) comparing to the ones exposed to a 3% humidification (Figure 6 a)), even if the exposure time is one order of magnitude higher in this case.

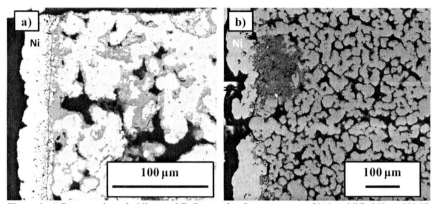

Figure 6. a) Cross section of a Ni coated FeCr sample after exposure to 3% humidified H_2 at 800 °C during 700 h. b) Cross section of a Ni coated FeCr sample after exposure to 50% humidified H_2 at 800 °C during 90 h.

Therefore, it is demonstrated that bare metal porous substrates are reasonably stable when operating at 50% humidified hydrogen fuel gas, whereas main cause of severe oxidation arises from the interaction between Ni and the FeCr from the metal substrate. This interaction diminishes when water vapour content of the fuel gas decreases or lower sintering temperatures are used during processing. However, interaction exists even at 800 °C which means that effective DBLs are necessary to stop the Ni-Fe-Cr interdiffusion.

In order to confirm these results and corroborate that the catastrophic oxidation observed in Figure 4 is caused by the deficient DBL employed to date (G1-type), two samples consisting of a porous metal support with the G1-type DBL, anode and nickel current collector, were deposited, sintered and later exposed to a 50% humidified hydrogen atmosphere during 70 and 300 hours respectively. These samples were run under 500 mA/cm^2 current load in order to further simulate fuel cell operating conditions.

Figure 7 shows the cross section of the three comparable samples, two of them tested for different exposure times to the highly humidified atmosphere: Figure 7 a) shows the microstructure of a sample after the sintering process at 1350 °C just before being exposed to the humidified atmosphere, while Figure 7 b) and Figure 7 c) show the microstructure of the samples after exposure to the humidified atmosphere for 70 and 300 hours respectively. Even after sintering in non-oxidizing atmosphere, relatively high amounts of Fe and Cr can be appreciated in the Ni particles of the anode and lower amounts of Ni in the ferritic alloy. After 70 hours exposure to the humidified atmosphere, elemental EDS analysis demonstrates the oxidation of Fe and Cr in the Ni particles of the anode, by the time that a thin protective chromia layer has grown in the metal support that still maintains the porous microstructure. However, after 300 hours under high water vapour atmosphere, microstructure completely changed and almost half of the metal support is completely oxidized. The presence of Fe and Cr in the anode and Ni in the ferritic alloy demonstrates that the G1-type DBL is not able to completely avoid interdiffusion between those elements.

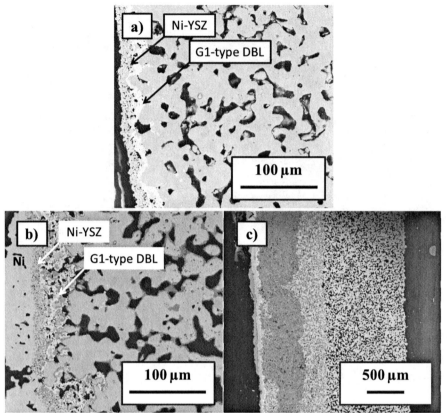

Figure 7. Cross section of the anodic part of the cell with G1-type DBL after different times of exposure to highly humidified (50% H₂O) atmosphere. a) t=0 h, b) t=70 h, c) t=300 h.

An alternative G2-type DBL was then proposed in order to avoid the interdiffusion of the elements that causes the degradation of the cell under simulated high fuel utilization conditions. This new DBL was used in equivalent stability tests to those described above at 800 °C during 500 hours in 50% humidified hydrogen under current loading of 500 mA/cm².

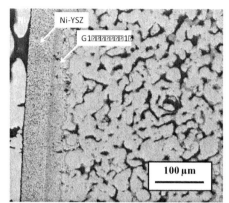

Figure 8. Cross section of the G2-type DBL containing anode and Ni current collector after 500 h exposure to 50% humidified hydrogen at 800 °C and 500 mA/cm^2.

A considerable improvement attributed to the new DBL can be observed in the micrograph shown in Figure 8 comparing to those in Figure 7. Moreover, the presence of Fe and Cr in the DBL decreases gradually as the distance from the Fe and Cr source increases, being almost negligible in the active anode layer and in the Ni current collector. Likewise, Ni was not detected in the ferritic alloy, so it is likely that cells fabricated using this alternative G2-type DBL should be resistant to oxidation when working under high fuel utilization conditions.

Performance of a Second Generation of SOFCs under High Fuel Utilization Conditions
Once the efficacy of the G2-type DBL in preventing the Ni, Fe and Cr interdiffusion has been demonstrated, full tubular metal-supported cells were fabricated and electrochemically tested at 800 °C under fuel utilization conditions similar to the ones used in first generation cells (excess gas flow of 3% humidified H$_2$). The resulting power density is considerably reduced for this type of cells due to the increase in polarization losses caused by lower porosity microstructures. Nevertheless, as demonstrated in Figure 9, durability under higher fuel utilization conditions was significantly improved. While dramatic degradation occurred after 13 hours when 5.6 ml/min·cm^2 of 3% humidified hydrogen were used in of first generation cells, the improved second generation cell performance remained almost unaltered for more than 650 hours under similar operating conditions. Moreover, the same cell was subsequently exposed to a simulated higher fuel utilization conditions (50% humidified hydrogen; FU = 50%) during another 100 hours, and although the applied current load had to be lowered to set the voltage value to 0.7 V, which lowered the power density from 125 to 81 mW/cm^2, no further degradation was observed.

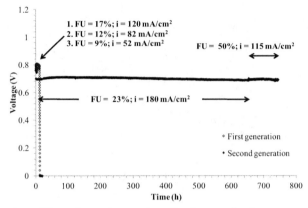

Figure 9. Comparison of first and second generation cells when tested under constant current load and real 17, 12 and 9% and 23% fuel utilization conditions respectively for the first 650 hours and highly humidified atmosphere (simulated 50% fuel utilization) for the last 100 hours at 800 °C.

CONCLUSIONS

First generation of tubular metal-supported SOFCs able to work under current loading for more than 2000 hours when 3% humidified hydrogen is fed in excess were successfully developed in Ikerlan. Resistance against thermal cycling have been demonstrated for more than 500 cycles, however, severe degradation is observed when fuel utilization conditions become more demanding. SEM/EDS analysis indicated that the main reason for the catastrophic degradation observed is the fact that the diffusion barrier layer is not able to avoid completely the interdiffusion between the Ni present in the anode and the Fe and Cr from the metal support. This point was analysed in detail with several tests carried out with the anodic components of the cell exposed to 50% humidified hydrogen. By replacing the former used diffusion barrier layer by an alternative one, more than 500 hours without significant degradation were achieved for the anode side components under 50%H_2O/50%H_2 at 800 °C. Moreover, a second generation of cells stable for more than 700 hours under high fuel utilization conditions has been successfully developed. Further work is currently devoted to improve cell performances and evaluation of thermal cycling capabilities of the new promising second generation cells.

ACKNOWLEDGEMENT

This work was founded by the PSE programme of the Spanish Government. Large equipment acquisition has been supported by FEDER funds from the European Union. Ikerlan wishes to thank CIC EnergiGUNE for loaning the Quanta FEG SEM.

REFERENCES

[1] EG&G Technical Services, Fuel Cell Handbook *U.S. Department of Energy*, Morgantown, (2004).
[2] M. C. Tucker, Progress in metal-supported solid oxide fuel cells: A review, *Journal of Power Sources,* **195**, 4570-82 (2010).
[3] M. Brandner, M. Bram, J. Froitzheim, H. P. Buchkremer and D. Stöver, Electrically Conductive Diffusion barrier layers for Metal-Supported SOFC, *Solid State Ionics,* **179**, 1501-04 (2008).

[4]J. Froitzheim, L. Niewolak, M. Brandner, L. Singheiser and W. J. Quadakkers, Anode side diffusion barrier coating for solid oxide fuel cells interconnects, *Journal of Fuel Cell Science and Technology*, **7**, 0310201-07 (2010).

[5]P. Szabo, J. Arnold, T. Franco, M. Gindrat, A. Refke, A. Zagst and A. Ansar, Progress in the metal supported solid oxide fuel cells and stacks for APU, *ECS transactions*, **25**, 175-85 (2009).

[6]P. Szabo, A. Ansar, T. Franco, M. Gindrat and A. Zagst, Dynamic electrochemical behaviour of metal-supported plasma-sprayed SOFC, *European Fuel Cell Forum Proceedings*, 16.44-16.53 (2010).

[7]D. Hathiramani, R. Vaßen, J. Mertens, D. Sebold, V. A. C. Haanappel and D. Stöver, Degradation mechanism of metal supported atmospheric plasma sprayed solid oxide fuel cells, *Ceramic Engineering and Science Proceedings*, **27**, 55-65 (2006).

[8]S. Hui, D. Yang, Z. Wang, S. Yick, C. Decès-Petit, W. Qu, A. Tuck, R. Maric and D. Ghosh, Metal-supported solid oxide fuel cell operated at 400-600 °C, *Journal of Power Sources*, **167**, 336-39 (2007).

[9]N. P. Brandon, D. Corcoran, D. Cummins, A. Duckett, K. El-Khoury, D. Haigh, R. Leah, G. Lewis, N. Maynard, T. McColm, R. Trezona, A. Selcuk and M. Schmidt, Development of metal supported solid oxide fuel cells for operation at 500-600°C, *Journal of Materials Engineering and Performance*, **13**, 253-56 (2004).

[10]M. C. Tucker, G. Y. Lau, C. P. Jacobson, L. C. DeJonghe and S. J. Visco, Stability and robustness of metal-supported SOFCs, *Journal of Power Sources*, **175**, 447-51 (2008).

[11]M. C. Tucker, G. Y. Lau, C. P. Jacobson, L. C. DeJonghe and S. J. Visco, Performance of metal-supported SOFCs with infiltrated electrodes, *Journal of Power Sources*, **171**, 477-82 (2007).

[12]T. Klemensø, J. Nielsen, P. Blennow, A. H. Persson, T. Stegk, B. H. Christensen and S. Sønderby, High performance metal-supported solid oxide fuel cells with Gd-doped ceria barrier layers, *Journal of Power Sources*, **196**, 9459-66 (2011).

[13]P. Blennow, J. Hjelm, T. Klemensø, Å. H. Persson, S. Ramousse and M. Mogensen, Planar metal-supported SOFC with novel cermet anode, *Fuel Cells*, **11**, 661-68 (2011).

[14]I. Villarreal, M. Rivas, L. M. Rodriguez-Martinez, L. Otaegi, A. Zabala, N. Gomez, M. A. Alvarez, I. Antepara, N. Arizmendiarrieta, J. Manzanedo, M. Olave, A. Urriolabeitia, N. Burgos, F. Castro and A. Laresgoiti, Tubular metal supported SOFC development for domestic power generation, *ECS transactions*, **25**, 689-94 (2009).

[15]M. Brandner, M. Bram, D. Sebold, S. Uhlenbruck, S. T. Ertl, T. Höfler, H. P. Buchkremer and D. Stöver, Inhibition of diffusion between metallic substrates and Ni-YSZ anodes during sintering, *Proceedings - Electrochemical Society*, **PV 2005-07**, 1235-43 (2005).

[16]Fuel cell Testing Network, FCTESTnet Procedures, ENK5-CT-2002-20657.

[17]L. M. Rodriguez-Martinez, L. Otaegi, M. A. Alvarez, M. Rivas, N. Gomez, A. Zabala, N. Arizmendiarrieta, I. Antepara, M. Olave, A. Urriolabeitia, I. Villarreal and A. Laresgoiti, Degradation studies on tubular metal supported SOFC, *ECS Transactions*, **25**, 745-52 (2009).

[18]L. M. Rodriguez-Martinez, L. Otaegi, E. Sarasketa, N. Gomez, N. Arizmendiarrieta, M. A. Alvarez, M. Rivas, N. Burgos, F. Castro, I. Villarreal and A. Laresgoiti, Influence of interconnects in long term stability of tubular metal supported SOFCs, *European Fuel Cell Forum Proceedings*, 16.34-16.43 (2010).

HIGHLIGHTING DOE EERE EFFORTS FOR THE DEVELOPMENT OF SOFC SYSTEMS FOR APU AND STATIONARY APPLICATIONS

David R. Peterson, Jacob S. Spendelow, and Dimitrios C. Papageorgopoulos
U.S. Department of Energy, Office of Energy Efficiency and Renewable Energy

ABSTRACT

The U.S Department of Energy's (DOE) Office of Energy Efficiency and Renewable Energy (EERE) supports a broad range of fuel cell research, development, and demonstration (RD&D) activities, including RD&D on truck auxiliary power units (APU) and stationary applications. While a portfolio of fuel cell technologies for these applications is supported, this communication will focus on solid oxide fuel cell (SOFC) development for APUs, combined heat and power (CHP), and energy storage (reversible fuel cells). Challenges, strategy, and progress toward meeting EERE market-driven technical and cost targets will be highlighted. Recent research, development, and demonstration achievements for SOFC activities ranging from materials R&D to field demonstrations also will be provided.

INTRODUCTION

The Fuel Cell Technologies (FCT) Program, within the US Department of Energy's (DOE) Office of Energy Efficiency and Renewable Energy (EERE), represents a comprehensive portfolio of activities that address the full range of technological and non-technological barriers facing the development and deployment of fuel cell and hydrogen technologies. The Program conducts focused efforts to enable widespread commercialization of fuel cell and hydrogen technologies in diverse sectors of the economy. In particular, the Program pursues a technology neutral approach aimed at reducing cost and improving both durability and performance of fuel cell systems for a range of applications. A detailed description of the Program, including technical and cost targets, can be found in the Multi-Year Research, Development and Demonstration Plan.[1]

The Program's expanded focus includes near-term fuel cell applications, such as stationary prime power, combined heat and power, backup power, and auxiliary power units, as well as long-term transportation applications. These applications represent opportunities to benefit from the inherent advantages of fuel cells (e.g., energy efficiency and fuel flexibility), especially high temperature fuel cells, to lead to significant energy savings and reduction of greenhouse gas and criteria pollutant emissions. They also represent near-term opportunities for fuel cell commercialization, inducing growth in the domestic fuel cell manufacturing and supplier base.

While performance requirements of fuel cells deployed in APU and stationary applications differ from transportation applications, some of the technical challenges are the same. All of these applications require that fuel cell technologies exhibit performance, durability, and lifecycle cost competitive with the incumbent technology and other alternative technologies, or provide other enhanced capabilities.

Efforts at EERE supporting the development and deployment of fuel cell systems for APU and stationary applications, with emphasis on solid oxide fuel cells (SOFCs), will be the focus of this communication. Specific R&D activities in support of these efforts will be described along with progress toward meeting EERE technical and cost targets.

AUXILIARY POWER UNIT FUEL CELL SYSTEMS

The combustion engines used to provide motive power in transportation systems are typically also used to provide electrical power for parasitic loads, including onboard entertainment and HVAC. This traditional approach carries a number of disadvantages, including low fuel efficiency, high emissions of greenhouse gases (GHGs) and criteria pollutants, noisy operation, and engine wear. Therefore, the introduction of fuel cell APUs has attracted significant interest for trucking, marine, recreational vehicle, and aviation applications. For the approximately 500,000 long-haul Class 7 and Class 8 trucks in the United States, emissions during overnight idling have been estimated to be 10.9 million tons of CO_2 and 190,000 tons of NO_x annually.[ii] The use of APUs for Class 7–8 heavy trucks to avoid overnight idling of diesel engines could save up to 280 million gallons of fuel per year and avoid more than 92,000 tons of NO_x emissions.[iii,iv]

As of July 2011, anti-idling legislation exists in all or part of 30 states,[v] further contributing to demand for low-cost, high-efficiency truck APUs. Truck APU fuel cells are expected to use diesel fuel to power environmental controls and peripheral electrical devices. Fuel cell APU systems could supply the same electrical source as conventional internal combustion engine-based APUs, which is a primary competition to the fuel cell-based APU. Some potential benefits of fuel cell-based APUs include an increase in fuel efficiency of as much as 40-50%, ability to meet current emissions standards without aftertreatment, and very low noise. Potential advantages of diesel-fueled APU systems based on high temperature fuel cells (e.g SOFCs) as opposed to low temperature fuel cells (e.g. PEMFCs) include operation at an increased system efficiency and simplification of the overall system design. These advantages are primarily related to the high temperature fuel cell's ability to operate on high CO content syngas fuels which are produced via steam reformation and/or partial oxidation of hydrocarbon-based fuels.

Delphi Automotive Systems, LLC is currently partnering with PACCAR and TDA Research, Inc. in an EERE-supported effort under the 2009 American Recovery and Reinvestment Act (ARRA) to design, develop, and demonstrate a 3-5kW SOFC APU capable of operating on ultra low sulfur diesel (ULSD) fuel for heavy duty commercial Class 8 trucks[vi]. Following a series of in-house tests, including on-vehicle testing, to validate the 'road worthiness' of the diesel APU, it will be installed on a high visibility fleet vehicle for a one year field demonstration of providing power for vehicle hotel loads and other needs under real-world operating conditions.

Over the duration of the current project, Delphi has advanced from its A-level system design to the current B-level design. Figure 1 compares the A-level and B-level designs. Some advances in going to the B-level system design include an increased power output, smaller package size, reduced mass, and subsystems capable of high volume manufacturing. Also, the B level design moves from two fuel cell stacks to a single one with a significantly larger cell active area (~400 cm^2).

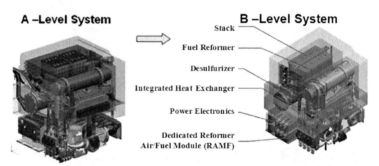

Figure 1. Comparison of Delphi A-Level and B-Level SOFC APU System Designs

Following the design and testing of the B-level subsystems, B-level systems have since been built and have undergone progressively more complex and realistic testing to ensure that the system is "road worthy." In January 2012, a B-level system will be installed on a truck for a one year demonstration which will consist of six months of testing in the northeast during cold weather and six months of testing in the south during hot weather.

Reduction of fuel consumption is a major driver of the effort to equip long-haul heavy-duty trucks with fuel cell APUs. Annual fuel savings resulting from incorporation of a fuel cell APU in a typical long-haul sleeper truck are projected to be on the order of 600 gallons/year.[vii] Realization of these fuel savings requires high-efficiency APU operation, with a long-term (2020) target of 40% electrical efficiency.[viii] The efficiency status currently recognized by EERE is 25%, which has been achieved by Delphi.[ix] In other EERE-funded work, completed in 2009, Cummins demonstrated a 4-module SOFC APU system (Figure 2) that achieved a DC efficiency of 11%.[x] The efficiency of the Cummins system was constrained by mismatch of component properties with system parameters. With additional system integration efforts, the Cummins system would be expected to produce markedly higher efficiency.

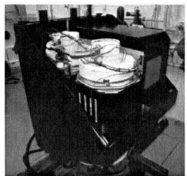

Figure 2. Cummins' assembled fuel cell modules and BOP components.[xi]

High power density and specific power are required to minimize size and weight to a level comparable to that of conventional (engine-based) APUs. Specific power of 45 W/kg and power density of 40 W/L are targeted by 2020.[viii] Current status is 20 W/kg and 17 W/L, respectively, based on results reported in response to a Request for Information (RFI) released in 2009, as well as more recent reports from EERE-sponsored projects.[viii]

High APU durability is required to maximize operational lifetime and minimize life-cycle cost, with a target lifetime of 20,000 hours identified as a level at which fuel cell APUs would be market-competitive. Existing EERE projects have not yet amassed sufficient operating hours to estimate fuel cell APU durability status, though limited demonstration of durability has been achieved. In work with the older A-level design, Delphi SOFC APU hardware was mounted on a Peterbilt Class 8 truck and driven 3000+ highway and secondary road miles. In the laboratory, the APU completed >50 thermal cycles. The stack has been tested to the equivalent of 3.5 million miles on a vibration table and the system has been tested to the equivalent of 17,000 highway miles on a vibration table.

COMBINED HEAT AND POWER FUEL CELL SYSTEMS

Fuel cells offer a highly efficient and fuel-flexible technology for distributed power generation and Combined Heat and Power (CHP) systems. While EERE supports small to mid-size fuel cell systems (up to several MWs), involving a 'technology neutral' stance where PEMFC, SOFC, and other fuel cell technologies are included, DOE's Office of Fossil Energy (FE) develops large-scale solid-oxide fuel cell systems for utility-scale distributed generation through the Solid State Energy Conversion Alliance (SECA) Program. Fuel cells have unique advantages in CHP applications. Currently in the U.S., 63 percent (or about 26 quadrillion BTU/year) of the total energy consumed for power generation is lost in the form of waste heat.[xii] The vast majority of this energy loss occurs at centralized power generation facilities. By distributing electrical power generation to sites with concurrent thermal energy needs, including homes and businesses, heat that would otherwise be lost may be recovered, thereby reducing total energy consumption.

Fuel cells are uniquely suitable for many commercial and residential applications, due to their quiet operation, ability to use existing natural gas fuel supplies, low operation and maintenance requirements, and ability to maintain high efficiency over a wide range of loads. High-temperature fuel cells such as SOFCs possess the advantage of producing high quality heat. Lower-temperature fuel cells, such as PEMFCs, are also being investigated for CHP applications, and are already being deployed for this purpose.[xiii,xiv]

EERE maintains a portfolio of distributed energy and CHP RD&D projects, including activities in RD&D of SOFCs, high-temperature PEMFCs, and low-temperature PEMFCs for micro-CHP applications. In the near future, plans exist to expand efforts to include medium scale CHP (e.g. greater than 100 kW but less than multi-MW.)

EERE is currently supporting efforts at Acumentrics Corporation to develop a low cost, 3-10 kW tubular SOFC power system. The Acumentrics system is based on a Ni/YSZ anode tube supporting a YSZ electrolyte and LSM or LSM/YSZ cathode, and operates through internal reforming of natural gas. Improvements in performance and durability have enabled an effort to commercialize, in partnership with the Ariston Thermal Group of Italy, a 1 kW wall-mounted micro-CHP unit for European markets.[xv] Reduction in manufacturing cost is a major component of Acumentrics' system cost reduction strategy. Increasing automation of cell manufacturing, along with incorporation of higher throughput manufacturing processes, continues to decrease cost. A 24% gain in power density since 2009 allowed Acumentrics to decrease the number of tubes per stack by 55%, leading to a 33% reduction in stack volume and a 15% reduction in stack weight, as depicted in Figure 3.[xvi] Durability is also improving, with over 12,000 hours operation demonstrated in 2011, more than double the 2010 demonstrated durability. These advancements, along with reduction in cost of cell components and further improvements in performance and durability, provide a path towards widespread commercialization of Acumentrics' tubular SOFC technology for micro-CHP applications.

Figure 3. Acumentrics decreased stack volume and weight.[16]

In addition to the SOFC development underway at Acumentrics, EERE is also supporting development of high-temperature PEMFCs for CHP applications at Plug Power, and low-temperature PEMFCs at Intelligent Energy. This communication is limited to discussion of SOFC RD&D, but details of the non-SOFC projects are available elsewhere.[xvii,xviii]

The high energy efficiency attainable with fuel cell CHP systems makes them attractive from an energy conservation and GHG emission reduction standpoint. EERE has established long-term electrical efficiency and CHP efficiency targets of 45% and 90%, respectively, in micro CHP (1 – 10 kW) applications, and 50% and 90%, respectively, in medium scale (100 kW – 3 MW) CHP applications. CHP efficiency in this case is defined as the sum of net regulated AC electrical energy plus recoverable thermal energy, divided by the lower heating value (LHV) of the fuel. Acumentrics has reported electrical efficiency up to 40% and CHP efficiency up to 85% with a tubular SOFC micro CHP system. While EERE-funded projects are making progress toward meeting efficiency targets, further work is needed to develop systems that can meet efficiency targets and other technical and cost targets concurrently.

EERE has established long-term cost targets at levels projected to allow widespread market penetration. For medium-scale CHP applications with rated power between 100 kW and 3 MW, equipment cost of $1000/kW for systems operating on natural gas is targeted in 2020, while $1400/kW is targeted for systems operating on biogas. Current status is estimated to range from $2500-$4500/kW and $4500-$6500/kW for systems operating on natural gas and on biogas, respectively.[xix]

For micro CHP applications, long term cost targets vary with system capacity, compensating for variation in installation costs. Equipment cost targets have been established for three levels of power (defined in this case as average power produced): $1,000/kW_{avg}, $1500/kW_{avg}, and $1700/kW_{avg} for 2-kW, 5-kW, and 10-kW systems, respectively.[xx] An analysis of modelled systems by Strategic Analysis, Inc. projects that status of 5-kW PEMFC micro CHP units based on current technology would be $2300-$4000/kW, depending on manufacturing volume.[xxi] Two new competitively selected independent cost analysis projects were awarded in 2011. The analyses, to be conducted by Lawrence Berkeley National Laboratory (LBNL) and Battelle Material Institute, will provide projections for both micro- and medium-scale CHP systems, at variable production volumes, for a variety of technologies including SOFCs.

Assumptions about required start-up time vary widely, and are strongly dependent on size and intended operating strategy of a CHP unit. EERE has established a 2020 target of 20 minutes, representing a value that is expected to meet consumer expectations without unduly increasing system cost and complexity. Acumentrics has reported start-up of their tubular SOFC system to full power in 20 minutes.

Based on their experience with highly-durable conventional household furnaces and boilers, consumers will expect fuel cell CHP systems to last for many years. EERE has established a 2020 target of 60,000 hours (approximately seven years) durability on a micro CHP operating cycle, with less than 0.3% degradation in performance per 1000 hours.[xx] With current durability status at 12,000 hours (Acumentrics), significant R&D work is required to increase lifetime. For medium scale CHP applications, likely to be installed at commercial sites including grocery stores, hotels, and office buildings, higher levels of durability are expected, with a 2020 target of 80,000 hours. However, the more mature technologies being deployed in medium scale applications already have relatively high durability, with 40,000 hours durability reported for molten carbonate fuel cells and 80,000 hours reported for phosphoric acid fuel cells.[xix]

REVERSIBLE FUEL CELLS

Reversible fuel cells, which are capable of operating in either fuel cell or electrolyzer mode, are of interest for energy storage applications, and hold promise as an enabler for implementation of intermittent renewable energy technologies. This technology allows for storage of excess energy during periods of low electricity demand, and deployment of the stored energy during times of peak demand.

High round-trip efficiency is a key requirement of a reversible fuel cell, requiring low-overpotential operation in both fuel cell and electrolyzer mode. SOFCs, with their extremely high electrical efficiency, are therefore a promising candidate for application as reversible fuel cells.

The National Renewable Energy Laboratory (NREL) and the DOE recently held a reversible fuel cells workshop which included a session on SOFCs and solid oxide electrolyzer cells (SOECs)[xxii]. The workshop participants recommended R&D to develop more robust materials, mitigate sources of degradation, and to perform techno-economic studies of complete SOFC/SOEC systems in renewable electricity storage applications.

EERE-supported R&D at Versa Power Systems (VPS), leveraging technology developed with the support of the SECA program, has resulted in significant progress in the optimization of SOFC technology for reversible operation[xxiii]. The VPS project emphasized improvement in reversible SOFC performance and durability through identification and characterization of degradation mechanisms, development of cell and interconnect materials, and stack design and demonstration. Reversible SOFCs developed at VPS have demonstrated over 2500 hours operation with daily cycling between fuel cell mode and electrolyzer mode, and with only 22 mV degradation in fuel cell voltage, corresponding to a degradation rate of 0.89% per 1000 hours calculated from fuel cell mode (see Figure 4). Reversible SOFCs developed through the EERE project have demonstrated operating current density as high as 500 mA/cm^2

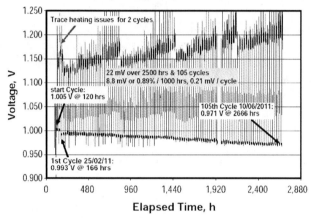

Figure 4. Fuel cell and electrolysis cycling testing of a Versa Power Systems reversible SOFC.[23]

Based on the latest information from VPS, NREL will conduct analysis of unitized SOFC/SOEC systems for energy storage applications. NREL will define product requirements, develop two energy storage scenarios, and perform lifecycle cost analyses for the systems using financial assumptions and methodology consistent with a previous study, which looked at a MW-scale alkaline electrolyzer/PEM fuel cell system.[xxiv]

CONCLUSION

EERE maintains a diverse portfolio of RD&D projects in fuel cell technologies for stationary and APU applications. These projects contribute to EERE's efforts to enable widespread commercialization of fuel cells through minimization of technical, institutional, and market barriers. Commercialization of fuel cell stationary and APU systems would increase energy savings and reduce emissions of greenhouse gases and criteria pollutants. Furthermore, deployment of fuel cells in these applications would help to build a robust supplier base to facilitate further commercialization efforts.

Progress of EERE-funded R&D work toward meeting technical targets has been described above, but significant further investment is required to achieve technical targets and overcome barriers to commercialization. Through continuation of its support for fuel cell APU and stationary RD&D efforts, EERE will continue to work toward widespread commercialization of fuel cell technologies in these applications.

ACKNOWLEDGMENT
Sunita Satyapal, Stephanie Byham, Norman Bessette, Gary Blake, Dan Norrick, Diane Aagesen, Brian James, Rick Cutright, and Randy Petri are gratefully acknowledged for their valuable contributions.

REFERENCES

[i] "Multi-Year Research, Development and Demonstration Plan," Fuel Cell Technologies Program, http://www1.eere.energy.gov/hydrogenandfuelcells/mypp/index.html

[ii] Nicholas Lutsey, Christie-Joy Brodrick & Timothy Lipman, "Analysis of Potential Fuel Consumption and Emissions Reduction from Fuel Cell Auxiliary Power Units (APUs) in Long Haul Trucks," Elsevier Science Direct, Energy 32, September 2005.

[iii] L. Gaines and C. Hartman, "Energy Use and Emissions Comparison of Idling Reduction Options for Heavy-Duty Diesel Trucks," Center for Transportation Research, Argonne National Laboratory, November 2008

[iv] Idle Reduction Technology: Fleet Preferences Survey, American Transportation Research Institute, February 2006.

[v] "Compendium of Idling Regulations," American Transportation Research Institute, http://www.atri-online.org/research/idling/ATRI_Idling_Compendium.pdf

[vi] Andrew Rosenblatt (Delphi) et al., Solid oxide Fuel Cell Diesel Auxiliary Power Unit Demonstration," in: FY 2011 Progress Report for the DOE Hydrogen and Fuel Cells Program, U.S. Department of Energy, Washington, DC, 2011, pp. 1297. http://www.hydrogen.energy.gov/pdfs/progress11/xii_4_rosenblatt_2011.pdf

[vii] DOE Hydrogen Program Record #9010, "Benefits of Fuel Cell APU on Trucks," November 2009 http://hydrogen.energy.gov/program_records.html

[viii] DOE Hydrogen Program Record #11001, "Revised APU Targets," January, 2011, http://hydrogen.energy.gov/program_records.html

[ix] Steven Shaffer (Delphi) et al., "Solid Oxide Fuel Cell Development for Auxiliary Power in Heavy Duty Vehicle Applications," in: FY 2010 Progress Report for the DOE Hydrogen Program, U.S. Department of Energy, Washington, DC, 2010, pp. 907. http://www.hydrogen.energy.gov/pdfs/progress10/v_i_2_shaffer.pdf

[x] Dan Norrick (Cummins) et al., "Diesel-Fueled SOFC System for Class 7/Class 8 On-Highway Truck Auxiliary Power," in: FY 2010 Progress Report for the DOE Hydrogen Program, U.S. Department of Energy, Washington, DC, 2010, pp. 903. http://www.hydrogen.energy.gov/pdfs/progress10/v_i_1_norrick.pdf

[xi] Charles J.P. Vesely III et al. (Cummins Power Generation), "Diesel Fueled SOFC for Class 7/Class 8 On-Highway Truck Auxiliary Power," Final Technical Report, March 31, 2010

[xii] Annual Energy Review, 2008. Energy Information Administration. Washington: June, 2009 http://www.eia.doe.gov/emeu/aer/pdf/aer.pdf

[xiii] ENE-FARM Program, http://www.ene-farm.info/en/

[xiv] ClearEdge Power, http://www.clearedgepower.com/

[xv] Norman Bessette (Acumentrics Corporation), "Development of a Low Cost 3-10 kW Tubular SOFC Power System," FY 2011 Progress Report for the DOE Hydrogen and Fuel Cells Program, U.S. Department of Energy, Washington, DC, 2011, pp. 890. http://www.hydrogen.energy.gov/pdfs/review11/fc032_bessette_2011_o.pdf

[xvi] Norman Bessette (Acumentrics Corporation), "Development of a Low Cost 3-10 kW Tubular SOFC Power System," FY 2010 Progress Report for the DOE Hydrogen Program, U.S. Department of Energy, Washington, DC, 2010, pp. 866. http://www.hydrogen.energy.gov/pdfs/review10/fc032_bessette_2010_o_web.pdf

[xvii] Donald Rohr (Plug Power), Highly Efficient, 5 kW CHP Fuel Cells Demonstrating Durability and Economic Value in Residential and Light Commercial Applications," in: *FY 2011 Progress Report for the DOE Hydrogen and Fuel Cells Program*, U.S. Department of Energy, Washington, DC, 2011, pp. 1325. http://www.hydrogen.energy.gov/pdfs/progress11/xii_12_rohr_2011.pdf

[xviii] Durai Swamy (Intelligent Energy) et al., "Development and Demonstration of a New Generation High Efficiency 10kW Stationary PEM Fuel Cell System," in: *FY 2011 Progress Report for the DOE Hydrogen and Fuel Cells Program*, U.S. Department of Energy, Washington, DC, 2011, pp. 894. http://www.hydrogen.energy.gov/pdfs/progress11/v_k_2_swamy_2011.pdf

[xix] DOE Hydrogen Program Record #11014, "Medium-Scale Fuel Cell System Targets," January 2012, http://hydrogen.energy.gov/program_records.html

[xx] DOE Hydrogen Program Record #11016, "Micro CHP Fuel Cell System Targets," January 2012, http://hydrogen.energy.gov/program_records.html

[xxi] Private communication with Brian James, Strategic Analysis, Inc., 2011.

[xxii] R.J. Remick and D.J. Wheeler, Reversible Fuel Cells Workshop Summary Report, (2011), http://www1.eere.energy.gov/hydrogenandfuelcells/wkshp_reversible_fc.html

[xxiii] Randy Petri (VPS) et. Al., "Advanced Materials for RSOFC Dual Operation with Low Degradation" in *FY 2011 Progress Report for the DOE Hydrogen and Fuel Cells Program*, U.S. Department of Energy, Washington, DC, 2011, pp. 876. http://www.hydrogen.energy.gov/pdfs/progress11/v_i_2_petri_2011.pdf

[xxiv] D. Steward, G. Saur, M. Penev, and T. Ramsden, NREL/TP-560-46719 (2009). http://205.254.148.40/hydrogenandfuelcells/pdfs/46719.pdf

EFFICIENT PLANAR SOFC TECHNOLOGY FOR A PORTABLE POWER GENERATOR

Andreas Poenicke, Sebastian Reuber, Christian Dosch, Stefan Megel, Mihails Kusnezoff, Christian Wunderlich, and Alexander Michaelis

Fraunhofer Institute for Ceramic Technologies and Systems IKTS
Winterbergstrasse 28, 01277 Dresden, Germany

ABSTRACT

Portable power generators for camping and industrial applications require start-up times around 30 minutes and need to achieve a life time of 3,000 h including 300 cycles. Preferably they operate on available fuels and have a compact and lightweight system design. Eneramic®, a portable solid oxide fuel cell (SOFC) system in the 100 Watt class has been developed. A planar SOFC stack based on electrolyte supported cells and ferritic interconnects is used for the eneramic® system. The long-term stability of SOFC stacks was tested over more than 3,000 hours with power degradation below 1.0 %/1,000 h. However, hotbox testing of 40-cell stacks and stack operation in the system environment revealed slightly higher degradation rates between 2.1 and 2.3 %/1,000 hours. SOFC stacks exposed to thermal cycles showed no power losses. The results show that the compact planar SOFC stack is capable to survive the expected system life time.

INTRODUCTION

The Fraunhofer Institute for Ceramic Technologies and Systems (IKTS) is developing fuel cell systems that are based on ceramic materials. In the past, several solid oxide fuel cell (SOFC) stack generations based on planar electrolyte supported cells (ESC) have been demonstrated[1,2]. Additionally, SOFC systems powered by fossil fuels with an electrical output ranging from 1 to 5 kW have been developed[3,4,5]. Under the brand name eneramic®, a micro-scale SOFC system with an electrical net output of 100 W has been designed at Fraunhofer IKTS[6].

The eneramic® SOFC system is intended as a remote power supply device for leisure, industrial and security applications. In these growing markets, portability, robustness and ease of use have a higher priority than long-term degradation, much in contrast to stationary applications. Thus a system concept was developed regarding the following design criteria: (1) use of existing fuels to solve the common problems of fuel availability, fuel safety and environmental issues; (2) water free system operation to avoid complicated water recirculation systems and freezing of the system at low-temperatures; (3) rapid cold start capability to reach start-up times below 30 minutes; (4) self-sustaining system operation with no need for an additional power or heat source; (5) net electrical system efficiency above 20 %; and (6) low cost system design which sustains 300 thermal cycles and 3,000 hours of operation by the use of commercially available actuators and sensors to minimize the costs for balance of plant (BoP) parts.

In the beginning of this work, the eneramic® system concept is introduced followed by a detailed description of the SOFC stack design. Afterwards, this work focuses on the characterization of the planar SOFC stack. The stack performance in furnace, in hotbox and in the system environment is emphasized, the power degradation during long-term operation is presented, and possible degradation mechanisms are discussed. However, the main challenges for the planar SOFC stack are transient operation modes and rapid start-up cycles. Thus, results on the thermal cycling of stacks in furnace are incorporated. Further results on cyclic stack operation in the system environment and on system efficiency can be found in detail in the work of Reuber et al.[6].

SYSTEM DESIGN AND MAJOR COMPONENTS

Many small fuel cell systems are limited in their fuel choice to proprietary fuel cartridges and fuels like hydrogen or methanol, which severely impacts refueling choices in remote areas. Of the fuel choices, only carbon based, energy rich fuels like propane and LPG (liquefied petroleum gas, a propane/butane mixture) are particularly attractive because (1) they are commercially available everywhere in the world; (2) they have a sufficiently high vapor pressure and can be easily supplied at ambient temperature to obviate the need for a fuel pump; (3) they can readily be stored under moderate pressure in liquid state with high energy densities; and (4) they impose no risk on the environment.

System Concept

Figure 1 shows the process flow diagram of the eneramic® system. The low pressure fuel is expanded over a flow control valve and purified to prevent harmful sulfur species from damaging reformer or stack. To operate the system without process water, the obvious choice is the use of a catalytic partial oxidation (CPOx) fuel processor. After the CPOx, the SOFC stack is fed by the reformate mixture and by pre-heated cathode air. The SOFC stack is operated at 850°C with a fuel utilization of 70 %. Downstream of the stack, all combustible residues remaining in the stack exhaust gas are completely oxidized in a porous afterburner. This afterburner is designed to maintain the carbon monoxide emission limits of the most well-known eco-label, the Blue Angel (50 mg/kWh) in all operation modes. For reasons of efficiency, the hot exhaust gas is used to heat the system core within the hotbox before it is fed to the cathode air pre-heater. Finally the exhaust is cooled down with ambient air to harmless temperatures below 50°C. In contrast to alternative approaches[7,8], a separate and modular start-up burner was integrated to accomplish a system cold start time of less than 30 minutes.

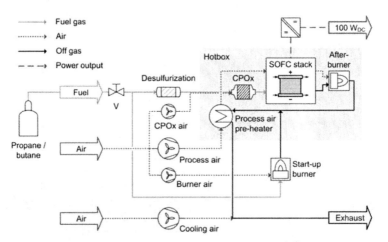

Figure 1. Simplified process flow diagram of the eneramic® system.

The core of the system consists of a single, mechanically compact unit comprising the CPOx, the SOFC stack, the afterburner, the process air pre-heater, and a central media distribution module (see Figure 2). The eneramic® system is designed without any piping, which helps to minimize pressure losses and heat dissipations. The use of multilayer technologies offers new design opportunities for internal gas manifolds. Therefore, all core components are tightly packaged in order to achieve a good thermal behavior of the system.

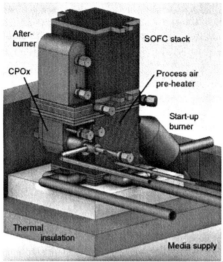

Figure 2. System core and main components of the eneramic® system.

Most of the core components are build-up from planar metallic parts. All metallic parts are essentially two dimensional and are manufactured from flat foils or coils. Thus, there is no need to apply deep drawing, welding or other cost-intensive technologies. Functional layers such as protective coatings, glass seals, or metallic brazes are applied using cost-effective multilayer technologies like screen printing or dispensing. Finally, the system components are manufactured by simply stacking the flat metallic parts followed by a joining step. The planar arrangement of the component parts was chosen because it offers the most cost-effective solution. Furthermore, due to the flat geometries the used multilayer technologies can be easily transferred into mass production.

Fuel Processing

The eneramic® system is designed for fossil fuels like propane, butane, or propane/butane mixture and for biofuels. However, before supplying the fuel to the fuel processor, it is purified in a separate removable reactor by extracting sulfur. This desulfurization cartridge is designed for quick and easy disassembling while changing the fuel cartridge. Odorants like amyl mercaptan and thiophene are absorbed by passing the fuel gas over a catalyzed fixed bed reactor.

The eneramic® CPOx reformer is tightly integrated into the system in terms of mechanical package, but also in terms of heat exchange with the other components. A well balanced heat management helps to keep the operation temperature at 800°C. The cylindrical reformer is equipped with a catalytically active honeycomb ceramic based on cordierite. The CPOx reformer operates at an efficiency of > 80 %. Regarding the reforming parameters, the risk of soot formation is relatively low. The average composition of the reforming gas was: 48 % N_2, 27 % H_2, 21 % CO, 2 % CO_2 and some water.

SOFC Stack

The planar SOFC stack is the most comprehensive and sensitive part of the eneramic® system which defines its electrical efficiency and power output. For small SOFC systems often tubular cells are applied as they sustain higher temperature gradients than planar cells[9,10,11]. However, the integration of tubular cells into a compact SOFC system is a difficult challenge. In contrast, planar electrolyte supported cells (ESC) offer good oxidation resistance and stability against redox and thermal cycles[12].

For the eneramic® system, a planar stack concept designed in cross-flow configuration with internal manifolds for reformate gas supply to the anode and air to the cathode is used. The SOFC stack is equipped with ESC based on a partly stabilized zirconia (3YSZ) electrolyte with a thickness of 90 µm in combination with an LSM/ScSZ cathode. The 3YSZ electrolyte is laser cut to integrate the manifolds and to ensure electric insulation between adjacent interconnects. During stack development the anode composition was improved. At first a Ni/8YSZ anode was used and later an anode made of Ni/GDC. The ESC are produced in-house at Fraunhofer IKTS with similar techniques used for electrolyte supported cells with high power density[13].

As interconnect material the ferritic steel Crofer22APU (trademark of ThyssenKrupp VDM) is selected due to its suitable thermal expansion, low electrical resistance, and high corrosion resistance. Planar SOFC stacks with Crofer22APU interconnect and ESC are a well-known and proved combination which can withstand thermal and redox cycles with minimal power losses[14,15]. Also the long-term stability of the materials for seals and interconnects at high temperatures is well documented[16,17].

Figure 3 shows the components of the SOFC stack used in the eneramic® system. One repeating unit consists of the ESC, the laser cut planar interconnect of Crofer22APU, two sealing frames, nickel meshes as anode contact, and ceramic ribs of LSMC as cathode contact. To prevent Cr evaporation and damage of the cathode a protective spinel coating is screen printed on the interconnect[18]. The sealing of interconnect and ESC is realized by a screen printed layer of a SiO_2-Al_2O_3-BaO sealing glass on both sides of a Crofer22APU frame. It may be added that during engineering of the stack design the main focus was set on the ease manufacturing of all essentially two-dimensional stack parts.

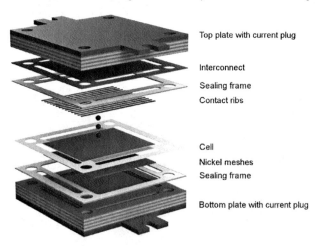

Top plate with current plug

Interconnect
Sealing frame
Contact ribs

Cell
Nickel meshes
Sealing frame

Bottom plate with current plug

Figure 3. Components of the eneramic® SOFC stack (exploded view).

Porous Media Reactors

In the eneramic[R] system two porous media reactors are integrated. The burner functionalities are generated with catalytically modified ceramic foam inserts. The ceramic foams are produced accordingly to a modified Schwartzwalder process at Fraunhofer IKTS[19]. The advantages of SiC are a high thermal conductivity, good mechanical strength, and well oxidation stability at high temperatures[20]. The SiC ceramic foams are placed in reactors which are located directly on the media distribution module. A start-up burner is used to heat-up the eneramic[R] system by combustion of propane or LPG.

In contrast to the fixed operating point of the start-up burner, the afterburner has to withstand rapid transients in gas quality during system start-up and abnormal system states such as sudden load shedding. For the realization of a stable combustion under all these different operating conditions the combustion in porous media is the best choice. It was shown that the afterburner can operate on very lean stack exhaust gas in normal system operation ensuring always a CO concentration below 50 mg/kWh in the exhaust gas. Thus, in all system states the CO concentration complies with the criteria of the eco-label Blue Angel.

Balance of Plant

The eneramic[R] system is designed without external piping between the central media distribution module and the other hotbox components, which helps to minimize pressure and heat losses. This allows for commercial off-the-shelf components for the balance of plant (BoP), e.g. the fluidic and electric management. Furthermore, the application of commercially available actuators and sensors helps to minimize the system costs even at an early stage of development.

Besides the costs, choosing the right BoP components helps to reach the desired system features: (1) to build a system with high gross efficiency, components with low power consumption should be used; (2) to maintain a reasonable buffer battery size, a start time up below 30 minutes should be reached; and (3) to sustain 300 thermal cycles and 3,000 hours of operation, the components should be long-term stable. Currently, a stack power output of at least 130 W_{DC} is necessary to provide 100 W_{DC} for the user and to power the parasitic BoP components.

Stack Characterization

Stacks were assembled by simply stacking the prepared components. Next, the assembled SOFC stacks were heated in a sealing furnace in air/nitrogen-flow with 2 K/min up to the sealing temperature of 900-930°C. After sealing, the stacks were cooled down to the 850°C. The nickel oxide in the anode was reduced step by step to metallic nickel by increasing the partial pressure of hydrogen in nitrogen. The gas tightness of the stack was evaluated by voltage of the cells at 0 A under a nitrogen/hydrogen gas flow. The cathode side was flushed with air.

Further stack characterization focused on long-term testing in a furnace or hotbox environment and on cycle testing under real-life conditions. For all tests, the stacks were fueled with simulated reformate, where the carbonic species were substituted through electrochemical equivalent amounts of H_2 or H_2O. The standard test temperature was 850°C.

In addition, SOFC stacks were tested in an eneramic[R] prototype system. The prototype system was fueled with various propane/butane fuel gas mixtures and heated-up by the start-up burner. After start-up at approximately 750°C stack temperature, the CPOx reformer is self-ignited at full load with an air ratio of 0.34. While raising the electrical current, the stack temperature grows up fast until the self-sustaining operation mode at 850°C and 4.6 A full load was reached.

RESULTS AND DISCUSSION

Stack Performance
  Currently, a single SOFC stack consisting of 40 electrolyte supported cells with 16 cm² active cell area and Ni/GDC anode produces at 850°C a maximum power output of 150 W at 0.7 V per cell on a synthetic reformate. The stack performance at various operating temperatures during furnace testing is shown in Figure 4.
  At a stack temperature of 765°C (measured at the top plate), a current density of 210 mA/cm² at 0.65 V per cell is measured. This is important, as the stack is operated at these low temperatures during start-up of the eneramic® system. Thereby, the CPOx is ignited when the stack temperature reaches 750°C. From this point the stack generates electricity from the CPOx reformate to suppress overheating of the afterburner and to accelerate the start-up process through electrochemical heat generation. While increasing the stack temperature to 815°C, the current density raises to 310 mA/cm² at 0.65 V per cell. After reaching 865°C, a maximum current density of 375 mA/cm² at the same cell voltage with a fuel utilization of 75 % is measured.

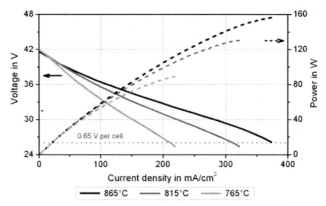

Figure 4. Current-voltage and current-power characteristics of a 40-cell SOFC stack with Ni/GDC anode running on simulated reformate ($H_2/N_2/H_2O$ = 48/48/4, V_{gas} = 280 sl/h) at various stack temperatures.

  Additionally, a well performing SOFC stack shows only small variation between the individual cell voltages after the activation procedure. For better comparison of the cell voltage distribution the area specific resistance (ASR) is calculated[21]. An example for the ASR distribution of two 40-cell SOFC stacks measured in a furnace is shown in Figure 5.
  In the first stack generation 3YSZ based ESC with Ni/8YSZ anode were implemented. Running a 40-cell stack on H_2/N_2 = 50/50 at 852 °C top plate temperature with 5 A (312 mA/cm²) yielded in ASR values between 0.65 and 0.75 $\Omega \cdot cm^2$. Further development at cell and stack level resulted in a new anode composition (Ni/GDC), in more accurate components of the stack, and in optimized height of the sealing glass. Through these efforts the mean variation of cell voltage and ASR could be lowered for the second stack generation. For the new ESC with Ni/GDC anode the stack ASR values are between 0.55 and 0.61 $\Omega \cdot cm^2$.

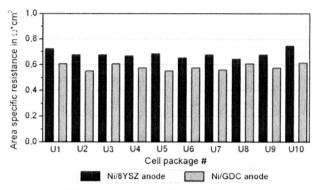

Figure 5. Comparison of ASR measurement of 40-cell SOFC stacks of first and second stack generation running on synthetic reformate ($H_2/N_2 = 50/50$).

Long-Term Stability

To determine the long-term stability of the SOFC stacks, a 5-cell stack was operated in a furnace for 3,200 hours. Figure 6 shows the time dependent behavior of the cell voltages of the 5-cell stack running on simulated reformate ($H_2/N_2/H_2O = 48/48/4$).

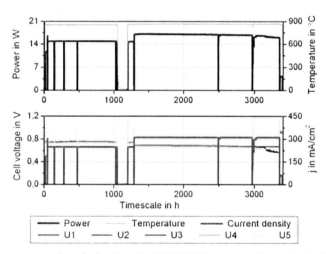

Figure 6. Long-term testing of a 5-cell stack with Ni/8YSZ anode running on simulated reformate ($H_2/N_2/H_2O = 48/48/4$) for more than 3,000 hours.

The long-term test started with an initial phase at 4 A (250 mA/cm^2) to activate the stack and to measure impedance spectra and current-voltage curves. During the first 260 hours, the cell voltages increase due to activation of the cells and a maximum power of $P_{el} = 14.9$ W was reached. From the long-term operation during the next 750 hours a power degradation of 0.93 %/1,000 h was calculated. Afterwards, a fixed operating point with the full load current of 5 A (312 mA/cm²) and a fuel utilization of 69 % was set and reached a power degradation of 0.89 %/1,000 h. No difference in degradation behavior was detected between the long-term operation at 4 A (250 mA/cm²) and at 5 A (312 mA/cm²). Unfortunately, after 3000 hours an emergency stop of the testing lab caused a rapid voltage drop of the first cell (U1). Simultaneously, an accelerated degradation of the cell voltage of the last cell (U5) was noticed.

The power loss in the stack was characterized using impedance spectroscopy. Figure 7 shows selected impedance spectra of cell 3 at initial state (after 24 h) and after 260, 1,300, and 3,000 hours. At initial state all cells show similar spectra starting with ohmic resistance R_0, followed by a small arch for the cathode polarization and a more distinct curve for the polarization at the anode. It was found that the contribution of the anode is different for all cells increasing from cell 1 up to cell 5. Regarding the spectra after 260 and 1,300 hours operation all cells show a reduction of the anode resistance due to further activation of the anode.

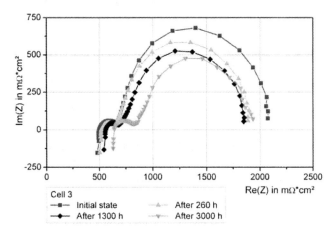

Figure 7. Selected impedance spectra of cell 3 of 5-cell stack with Ni/8YSZ anode during long-term testing at 850°C running on simulated reformate ($H_2/N_2/H_2O$ = 48/48/4).

After 3,000 hours operation the spectra exhibit several phenomena. The ohmic resistance R_0 for all cells changed: R_0 of cell 3 shows a slight increase, cells 2 and 4 show a similar behavior, R_0 of cell 1 is doubled and cell 5 reveals a dramatic increase of R_0 by factor 10 (not shown in Figure 7). The slight increase of R_0 of the middle cells can be attributed to an increase of the cathode resistance. A more detailed analysis revealed that an oxide layer was found at the interconnect surface after long-term operation, which is responsible for the increase of cathode resistance[21]. Noteworthy, the anode resistance has no effect on the power degradation due to ongoing activation during long-term operation.

Additional to the long-term characterization, the complete 5-cell stack with compression elements was embedded in a polymer resin and cut perpendicular to the contact ribs. A detailed analysis via electron microscopy revealed that the contact ribs of some cells lost contact to the cathode leading to the observed increase of the ohmic resistance. It was found that the half of all contact ribs of cell 5 and only one contact rib of cell 1 spelt off. All other cells showed hermetic contact between ribs and cathodes. In agreement with results gained on model samples, the detected increase of the cell voltages of cells 1 and 5 can be attributed to the loss of contact at the cathode[21].

Long-Term Stability in Hotbox

The operating conditions have a significant influence on the temperature distribution inside the SOFC stack and the resulting degradation rate. To verify differences between operation in a furnace and self-sustaining operation in a hotbox another long-term test was conducted in a hotbox. A 40-cell stack with Ni/8YSZ anode was operated with simulated reformate ($H_2/N_2/H_2O = 48/48/4$) as fuel for 2,200 hours. The stack was operated at a fuel utilization of 69 % and individual cell voltages (5 cells per package) were measured. After initialization, a stable operating point at a stack temperature of 859°C with $P_{el} = 140.2$ W was achieved, Figure 8.

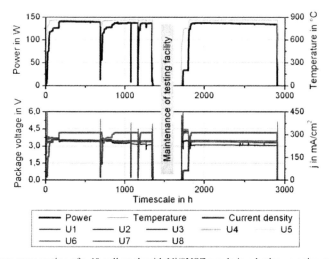

Figure 8. Long-term testing of a 40-cell stack with Ni/8YSZ anode in a hotbox running on simulated reformate ($H_2/N_2/H_2O = 48/48/4$) for 2,200 hours.

After long-term operation at a constant current of 5 A (312 mA/cm^2) and 4 thermal cycles due to maintenance stops, the power was decreased to $P_{el} = 133.8$ W. From the power loss an overall degradation rate of 2.1 %/1,000 hours was calculated which is slightly higher than the power degradation of the 5-cell stack measured in a furnace (Figure 6). The power loss can be attributed to the rapid degradation of the first cell package. However, a stack power output of at least 130 W_{DC} after 3,000 hours is still sufficient for the user and the parasitic BoP components over the lifetime of the eneramic® system.

Thermal Cycling

A 5-cell stack with Ni/8YSZ anode was subjected to thermal cycles with heating rates of 5 K/min in a furnace, Figure 9. After cycling, when the stack temperature reaches 850°C, the stack was operated at a constant current of 4 A (250 mA/cm^2) for 20 hours. Afterwards, the cell voltage distribution is always constant between 0.75 and 0.76 V. However, when reaching the 4 A operation point at first a decrease of the cell voltages is visible. During the following electricity generation the cell reactivate possible due to reversible contact losses at the anode. Additionally, with the planar SOFC stacks heating rates of 16 K/min were already realized in the system environment without any cell fractures or notable degradation losses.

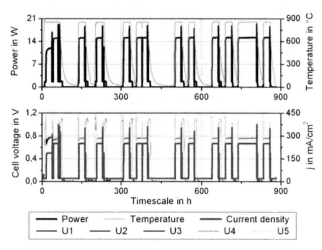

Figure 9. Cycle testing of a 5-cell stack with Ni/8YSZ anode in a furnace.

System Performance

For performance test in the system environment, the prototype system was equipped with a 40-cell SOFC stack with Ni/GDC anode. Figure 10 shows recorded data for 250 hours system operation with a propane fuel gas mixture. In stationary mode, the 40-cell SOFC stack achieves a power output of 133 W at a fuel utilization of 71 %. For the whole system a gross electrical efficiency of 31 % was calculated. The additionally shown temperatures of CPOx and afterburner demonstrate the temperature distribution during self-sustaining system operation. The stack temperature (measured at the top plate) is controlled at 859°C by varying the cathode air flow.

Individual cell voltages (4 cells per package) are measured, showing some degradation. The overall power degradation of the SOFC stack including all cells was calculated to 4.1 %/1000 hours. A detailed analysis of the recorded data revealed that the cell voltage of the upper cell packages U9-U10 drops constantly, probably due to insufficient stack compression. By excluding these cells from the calculation of a power degradation of 2.3 %/1000 h was achieved. This is in good agreement with the results from the long-term hotbox testing of 40-cell stacks (Figure 8).

The test proved that the SOFC stacks run properly on propane while integrated in the eneramic® system environment. In addition, variations on the propane/butane mixtures and stack operations with fuel utilization rates up to 82 % for more than 1,000 hours show little influence on stack performance[6].

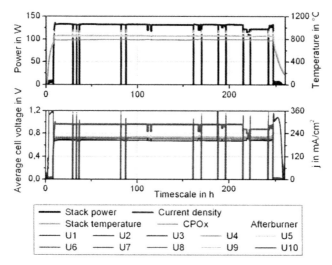

Figure 10. Operation of the eneramic® system with a 40-cell SOFC stack with Ni/GDC anode at 859°C stack temperature fueled with a propane fuel gas mixture.

CONCLUSION

The 100 W eneramic® system shows the potential of small scale SOFC systems for mobile and distributed power applications. A planar SOFC stack based on electrolyte supported 3YSZ cells laser cut sheet metal components and screen printed joining and functional layers can be produced with low costs. At 850°C a 40-cell SOFC stack with the newly developed Ni/GDC anode generates a power output of 150 W at 0.7 V per cell. During system operation a net electrical system efficiency of 31 % was reached with a similar stack. Degradation rates between 2.1 and 2.3 %/1000 h for 40-cell stacks with Ni/8YSZ anodes tested in a hotbox and in system context were determined from long-term testing under constant current. When using these degradation rates and regarding an initial power output of 150 W, the stack power will be still sufficient for the user and the parasitic BoP components over the lifetime of the eneramic® system. However, more important for the system durability are thermal cycles. The realized thermal cycle tests on 5-cell stacks showed no cell fractures or power losses.

Current research focuses on real-life testing regarding the long-term stability of the stacks with the Ni/GDC anodes and the ability for thermal cycles resulting from starts and stops of the system prototype. Furthermore, harsh gradients will be tested at system level. Besides, a more reasonable production of the cells and the other stack components will be pursued.

ACKNOWLEDGEMENTS

The Fraunhofer Future Foundation is gratefully acknowledged for funding of the eneramic® system development.

REFERENCES

[1]M. Kusnezoff, S. Megel, A. Paepcke, V. Sauchuk, A. Venskutonis, W. Kraussler, and M. Brandner, CFY-Stack Development for long-term Operation with high Efficiency, *Proc. 9th Eur. Solid Oxide Fuel Cell Forum*, Chap. 17, 9-19 (2010).

[2]S. Megel, M. Kusnezoff, N. Trofimenko, V. Sauchuk, A. Michaelis, A. Venskutonis, K. Rissbacher, W. Kraussler, M. Brandner, C. Bienert, and L.S. Sigl, High Efficiency CFY-Stack for High Power Applications, *ECS Trans.*, **35** [1], 269-77 (2011).

[3]M. Jahn, M. Stelter, M. Heddrich, E. Friedrich, and M. Kusnezoff, Development and Operating Experience with a SOFC-CHP System for Biogas, *Proc. 8th Eur. Solid Oxide Fuel Cell Forum*, B0403 (2008).

[4]M. Heddrich, M. Jahn, M. Stelter, and J. Paulus, Development of robust SOFC microCHP systems, *Proc. 9th Europ. Solid Oxide Fuel Cell Forum*, Chap. 3, 15-21 (2010).

[5]M. Heddrich, M. Jahn, K. Kaden, and A. Michaelis, Biogas SOFC µCHP – A simple process concept with high electrical efficiency, *Proc. 9th Eur. Solid Oxide Fuel Cell Forum*, Chap. 4, 88-95 (2010).

[6]S. Reuber, M. Schneider, M. Stelter, and A. Michaelis, Portable µ-SOFC System Based on Multilayer Technology, *ECS Trans.*, **35** [1], 251-8 (2011).

[7]C.M. Finnerty, C.R. Robinson, S.M. Andrews, Y. Du, P.M. Cheekatamarla, P.G. DeWald and T. Schwartz, Portable propane micro-tubular SOFC system development, *ECS Trans.*, **7** [1], 483-92 (2007).

[8]P.K. Cheekatamarla, C.M. Finnerty, C.R. Robinson, S.M. Andrews, J.A. Brodie, Y. Lu, and P.G. DeWald, Design, integration and demonstration of a 50W JP8/kerosene fueled portable SOFC power generator, *J. Power Sources*, **193** [2], 797-803 (2009).

[9]K. Kendall, Progress in Microtubular Solid Oxide Fuel Cells, *Int. J. Appl. Ceram. Technol.*, **7** [1], 1-9 (2010).

[10]D. Cui, Y. Du, K. Reifsnider, and F. Chen, One thousand-hour long term characteristics of a propane-fueled solid oxide fuel cell hot zone, *J. Power Sources*, **196** [15], 6293-8 (2011).

[11]Y. Du, D. Cui, and K. Reifsnider, Characterization of Propane-Fueled SOFC Portable Power Systems, *ECS Trans.*, **35** [1], 167-78 (2011).

[12]A. Glauche, T. Betz, S. Mosch, N. Trofimenko, and M. Kusnezoff, Long-term, Redox and Thermal Cycling Stability of Electrolyte Supported Cells, *ECS Trans.*, **25** [2], 411-9 (2009).

[13]N. Trofimenko, M. Kusnezoff, and A. Michaelis, Recent Development of Electrolyte Supported Cells with High Power Density, *ECS Trans.*, **35** [1], 315-25 (2011).

[14]B.E. Mai, T. Heller, D. Schimanke, J. Lawrence, and C. Wunderlich, Influence of Operating Conditions on the Reliable Performance of Stacks and Integrated Stack Modules, *ECS Trans.*, **25** [2], 187-94 (2009).

[15]J. Brabandt, Q. Fang, D. Schimanke, M. Heinrich, B.E. Mai, and C. Wunderlich, System Relevant Redox Cycling in SOFC Stacks, *ECS Trans.*, **35** [1], 243-9 (2011).

[16]J. Schilm, A. Rost, M. Kusnezoff, and A. Michaelis, Sealing Glasses For SOFC – Degradation Behaviour, *Ceram. Eng. Sci. Proc.*, **30** [4], 185-93 (2009).

[17]Q. Fang, M. Heinrich, and C. Wunderlich, CroFer22 APU as a SOFC interconnector material, *Proc. 9th Eur. Solid Oxide Fuel Cell Forum*, Chap. 12, 23-37 (2010).

[18]M. Kusnezoff, S. Megel, V. Sauchuk, E. Girdauskaite, W. Beckert, and A. Reinert, Impact of Protective and Contacting Layers on the Long-Term SOFC Operation, *Ceram. Eng. Sci. Proc.*, **30** [4], 83-93 (2009).

[19]J. Adler, G. Standke, M. Jahn, and F. Marschallek, Cellular Ceramics made of Silicon Carbide Ceramics for Burner Technology, *32nd Inter. Conf. Expo. Adv. Ceram. Comp.*, Daytona Beach, FL, USA, Jan. 27th – Feb. 1st, 2008.

[20]A. Fuessel, D. Boettge, J. Adler, F. Marschallek, and A. Michaelis, Cellular Ceramics in Combustion Environments, *Adv. Eng. Mater.*, **13** [11], 1008-14 (2011).

[21]S. Megel, Kathodische Kontaktierung in planaren Hochtemperatur-Brennstoffzellen, PhD Thesis, *Fraunhofer Verlag*, Stuttgart, Germany, ISBN 978-3-8396-0066-5, 2010.

INVESTIGATION OF NI-YTTRIA STABILIZED ZIRCONIA ANODE FOR SOLID-OXIDE FUEL CELL USING XAS ANALYSIS.

Koichi Hamamoto[1], Toshio Suzuki[1], Bo Liang[1], Toshiaki Yamaguchi[1], Hirofumi Sumi[1], Yoshinobu Fujishiro[1], Brian Ingram[2], A. Jeremy Kropf[2], and J. David Carter[2]

[1] National Institute of Advanced Industrial Science and Technology (AIST), Nagoya, Japan.
[2] Argonne National Laboratory, Argonne, IL, USA

ABSTRACT

Solid-oxide fuel cell (SOFC) anodes (Ni-Yttria stabilized ZrO_2) were investigated using X-ray absorption spectroscopy (XAS) after the fuel cell had operated at 650°C using H_2 fuel. The results show that Ni is oxidized up to 36 μm from the electrolyte and then becomes metallic nearer to the surface of the anode facing the reducing atmosphere. Using data from the X-ray absorption near edge structure (XANES) energy region, the metallic Ni fraction as a function of distance from the electrolyte has also been measured. XANES observation has shown the oxidation status of Ni in the anode is quite sensitive to position relative to the anode/electrolyte interface. These results may be an indication of the positions of the reaction sites for oxide ions and the fuel: assuming the oxidation state of nickel (Ni-O) designates the active triple phase boundary.

1. INTRODUCTION

Energy and environmental issues have never been as important world-wide as today. In addition to reducing CO_2 emissions, the efficient use of energy resources is critical. Fuel cells, known as environmental friendly power devices, are advantageous in this regard due to their high energy conversion efficiency. Among the variety of fuel cell types, the solid oxide fuel cell (SOFC) has been considered to be ideal as a future power source because it has the highest energy efficiency and potentially the longest operating lifetime. Current SOFC research topics focus mainly on lowering the operating temperature below 700 °C, increasing thermal shock resistance for quick start-up, and increasing the redox cycling stability. Such advancements can be achieved by developing new SOFC component materials and improving the electrode microstructures (6-11). In addition to these developments, the influence of the electrode microstructure on the cell performance is also revealed (12). Thus, a basic understanding of microstructural effect on the electrode performance becomes more important. Ni-cermet is typically used as anode, and changes its microstructure during the cell operation; Ni can be re-oxidized and reduced by transporting oxide ions. Thus it is of importance to understand microscopic Ni status in the anode. For this purpose, X-ray absorption spectroscopy was utilized to examine the fractured cross section of micro tubular SOFC anodes, and in this paper, we show the results of XANES analyses on the anode after the fuel cell operation.

2. EXPERIMENTAL

The micro tubular SOFC reported here consisted of nickel-yttria stabilized zirconia (Ni/YSZ) anodes, scandia stabilized zirconia (ScSZ) electrolytes, and $La_{0.6}Sr_{0.4}Co_{0.2}Fe_{0.8}O_3$-gadolinia doped ceria (LSCF/GDC) cathodes, with a GDC interlayer between the cathode and the electrolyte. Note that all of the materials used for the SOFC fabrication are commercially available.

Fabrication

Anode tubes were made from nickel oxide (NiO, Sumitomo Metal Mining Co., Ltd.), 8% yttria stabilized zirconia (YSZ, Tosoh Co., Ltd.), poly methyl methacrylate beads (PMMA, Sekisui Plastics Co., Ltd.), and cellulose (Yuken Kogyo Co., Ltd.). These powders were blended for 1h using a high-viscosity mixer (5DMV-r, Dalton Co., Ltd.); and, after adding the correct amount of water, were kneaded for 30 min under vacuum. The mixture was kept in a closed container over 15 h for aging. Tubes were extruded from the mixture using a metal mold (2.4 mm diameter with 2.0 mm diameter pin) by a piston-type extruder (Ishikawa-Toki Tekko-sho Co., Ltd.).

A slurry for dip-coating the electrolyte was prepared by mixing 10% scandium stabilized zirconia (ScSZ, Daiichiki-genso Co., Ltd.), solvents (toluene and ethanol), binder (poly vinyl butyral), dispersant (polymer of an amine system) and plasticizer (dioctyl phthalate) for 24 h. The anode tubes were dipped in the slurry and coated at the pulling rate of 1.0 mm s^{-1}. The coated films were dried in air, and co-sintered at 1350°C, for 1 h in air. An inter-layer of gadolinia doped ceria (GDC, Shinteu kagaku, Co., Ltd.) was dip-coated on the electrolyte layer of the tube and sintered at 1100 °C. Then, the inter-layer was dip-coated in a slurry containing $La_{0.6}Sr_{0.4}Co_{0.2}Fe_{0.8}O_{3-y}$ (LSCF) and GDC (LSCF/GDC). The SOFCs were completed by sintering at 1050 °C. The cell size was 1.8 mm diameter and 30 mm length with cathode length of 10 mm, having an effective electrode area of 0.56 cm^2.

Characterization

The microstructure of the electrodes of the tubular cell was characterized (before and after cell testing) using mercury porosimetry (Carlo Erba instruments, Pascal 140, 440) and scanning electron microscopy (SEM) (JEOL, JSM6330F). The cell was set in the furnace at a temperature of 650°C. Detailed information for experimental setup can be found in previous literature (13). During hydrogen was applied to the anode for 8 h at 650°C in the furnace, IV measurement was taken place every ten minutes and then the furnace was shut off. Hydrogen was kept flowing until the temperature had decreased below 200°C to avoid cell re-oxidation.

Microbeam XAS studies were performed at the Materials Research Collaborative Access Team (MR-CAT) beamline 10-ID at the Advanced Photon Source, Argonne National Laboratory (USA). A cryogenically cooled Si(111) monochromator selected the incident energy and a rhodium-coated mirror rejected higher order harmonics of the fundamental beam energy. The beam was focused to about a 2 μm x 2 μm spot using a platinum-coated KB mirror pair. All mirror angles were set with a critical energy near 10 keV so that higher order harmonics were attenuated by three reflections. The samples were positioned with the tube ends facing the beam with normal incidence to minimize the beam interaction with multiple layers in the tube. A four-element silicon detector (Vortex ME4, SII NanoTechnology USA Inc.) was used to detect the X-ray fluorescence. This detector was pointed toward the sample at an angle of about 40 degrees from perpendicular. This orientation resulted in a large self-absorption attenuation of the XAS amplitudes, especially for the highly-concentrated metallic nickel layer. The data was processed using Athena (14). The sample data was aligned using a Ni transmission standard placed to intercept elastically scattered x-rays from a kapton film. A self-absorption correction was made empirically with the "Fluo" algorithm in Athena (14) using the chemical formula: $(Ni)46(ZrO2)52(Y2O3)4$, angle in: $10°$ and angle out: $80°$ for the concentrated metallic nickel regions; and required different angles in: $20°$ and out: $15°$ for high NiO concentrated regions to obtain a good fit between the Ni and NiO standards. The relative concentrations of Ni and NiO in each sample were measured using a linear combination fitting algorithm with a fitting range of -20 and 30 eV below and above the edge using transmission Ni and NiO standards.

3. RESULTS AND DISCUSSION

Figure 1 shows cross-sectional fracture SEM images of micro tubular SOFC, of which the anode tube was sintered at 1350 °C. The final thickness of the electrolyte is ~6 μm. Figure 1 also shows the positions of SEM and XAFS analysis for the cell after fuel cell test. The cumulative pore-volume distribution (measured by mercury porosimetry) of the anode before reduction (oxide state) and after reduction are shown in Fig. 2. The dominant pore size of the anodes before reduction is 0.1 μm. After reduction, the distribution shifts to larger pore diameter and volume, due to the reduction from NiO to Ni, which led to the volume expansion. The pore volume of the cell increases from 0.05 to 0.09 $cm^3 g^{-1}$ and the porosity of the cell increased by 50% from 22 to 33% porosity. It is noteworthy that the formation of meso-pores in the range of 10 nm can also be identified in the cell.

The SEM images shown in Fig. 3 were taken (a) in the vicinity of the electrolyte, (b) at 40 μm and (c) at 100 μm from the electrolyte interface. XAS data was collected at 0, 2, 6, 16, 36 and 106 μm distances from the electrolyte interface. It is observed that the particles are agglomerated; however, the size of the particles was uniform. It is also seen that there is not much difference in the microstructure of each position.

On the other hand, XAS observation showed distinguishable chemical differences. In Fig. 4, the cell showed that Ni was oxidized up to 36 μm from the electrolyte and then became more metallic nearer to the surface of the anode facing the reducing atmosphere.

Figure 5 shows the Ni fraction as a function of distance from the electrolyte estimated from Fig. 4. Assuming that the oxidation state of Ni (Ni-O) indicates the position of the triple phase boundary, the results of the anode in Fig. 5 indicated that the anode had large number of possible reaction points ranging from the interface to 10~20 μm below the electrolyte.

Further XAFS investigation of the anode is necessary in order to gain the understanding and correct interpretation of the data, however, the results indicate that an XAS study has the potential to understand the electrochemical mechanism in SOFC anode and is considered to be excellent tool for further development of SOFC toward commercialization.

4. CONCLUSIONS

A conventional Ni-yttria stabilized zirconia (YSZ) solid-oxide fuel cell anode was investigated using XAS analysis. XANES observations have shown that the oxidation state of Ni in the anode is sensitive to position in the cross section, which may indicate the reaction sites (triple phase boundary) for oxide ions and the fuel. Further investigation is necessary for better understanding of the phenomena; however, it was shown that the XAS analysis can be a strong tool for fundamental investigation.

ACKNOWLEDGMENTS

This work is supported by Minister of Economy, Trade and Industry, Japan-U.S. cooperation project for research and standardization of Clean Energy Technologies. MRCAT operations are supported by the Department of Energy and the MRCAT member institutions.

REFERENCES

[1] N. Q. Minh, CERAMIC FUEL-CELLS. *J Am Cer Soc* **78**, 563-588 (1993).

[2] O. Yamamoto, Solid oxide fuel cells: fundamental aspects and prospects. *Electrochimica Acta,* **45**, 2423-2435. (2000)

[3] S. C. Singhal, Solid oxide fuel cells for stationary, mobile, and military applications. *Solid State Ionics,* **152–153**, 405-410 (2002) .

[4] B. C. H. Steele, A. Heinzel, Materials for fuel-cell technologies. *Nature,* **414**, 345-352, (2001).

[5] H. Yokokawa, N. Sakai, T. Horita, K. Yamaji, M. E. Brito, Electrolytes for solid-oxide fuel cells. *MRS BULLETIN* **30**, 591-595 (2005).

[6] J. W. Yan, H. Matsumoto, M. Enoki, T. Ishihara, High-power SOFC using $La_{0.9}Sr_{0.1}Ga_{0.8}Mg_{0.2}O_{3-\delta}/Ce_{0.8}Sm_{0.2}O_{2-\delta}$ composite film. *Electrochem Solid-Sate Lett* **8**, A389-A391 (2005) .

[7] Z. P. Shao, S. M. Haile, A high-performance cathode for the next generation of solid-oxide fuel cells. *Nature* **431**, 170-173(2004).

[8] S. W. Tao, J. T. S. Irvine, A redox-stable efficient anode for solid-oxide fuel cells. *Nature Materials* **2**, 320-323 (2003).

[9] K. Eguchi, T. Setoguchi, T. Inoue, H. Arai, Electrical-Properties of Ceria-Based Oxides and their Application to Solid Oxide Fuel-Cells. *Solid State Ionics,* **52**, 165-172 (1992).

[10] T. Hibino, A. Hashimoto, K. Asano, M. Yano, M. Suzuki, M. Sano, An intermediate-temperature solid oxide fuel cell providing higher performance with hydrocarbons than with hydrogen. *Electrochem Solid-Sate Lett ,* **5**, A242-A244 (2002).

[11]E. D. Wachsman, Functionally gradient bilayer oxide membranes and electrolytes. *Solid State Ionics* **152**, 657-662 (2002).

[12]T. Suzuki, M. H. Zahir, Y. Funahashi, T. Yamaguchi, Y. Fujishiro, M. Awano, Impact of Anode Microstructure on Solid Oxide Fuel Cells. *Science* **325**, 852-855(2009).

[13]T. Suzuki, T. Yamaguchi, Y. Fujishiro, and M. Awano, Improvement of SOFC Performance Using a Microtubular, Anode-Supported SOFC. *J. Electrochem. Soc.* **153**, A925-A928(2006).

[14]B. Ravel , Athena 0.8.054, accessed May, 18 2011 (2008) from:
http://cars9.uchicago.edu/ifeffit/Downloads.

Fig. 1: Fracture cross section SEM images of micro tubular SOFCs and the position of SEM and XAS analysis for the cell

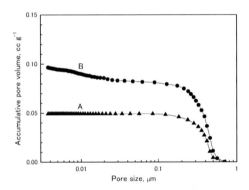

Fig. 2: Pore distribution of the anode, before (A) and after the reduction (B). The formation of meso pores become apparent after reduction (B).

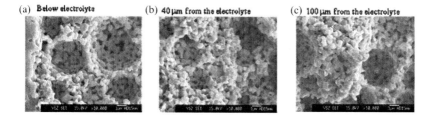

(a) Below electrolyte   (b) 40 μm from the electrolyte   (c) 100 μm from the electrolyte

Fig. 3: Fracture SEM images of the anode microstructures after fuel cell test at each position.

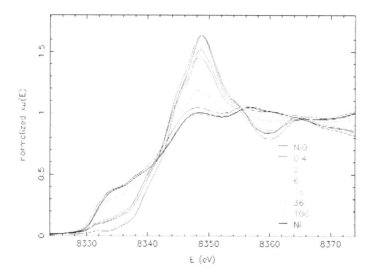

Fig. 4: XANES spectra of the anode obtained at each position of the fractured cross section of a previously tested cell.

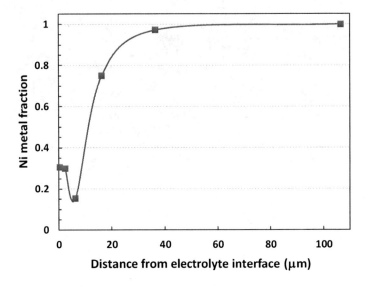

Fig. 5: Nickel metal fraction as a function of the distance from the electrolyte.

PROCESSING OF GADOLINIUM-DOPED CERIA ELECTROLYTE LAYERS WITH A THICKNESS OF ~1 MM: THIN FILM WET COATING METHODS AND PVD

Tim Van Gestel, Hyo-Jeong Moon, Doris Sebold, Sven Uhlenbruck, and Hans Peter Buchkremer

Forschungszentrum Jülich GmbH
Institute of Energy and Climate Research
IEK-1: Materials Synthesis and Processing
Jülich, Germany

ABSTRACT
    In this paper, two thin film processing methods are reported that deliver thin film GDC electrolyte layers. In the first part, the preparation of GDC layers with a wet-coating method is described. These layers were deposited in a similar way as conventional suspension based layers, but the essential difference includes the use of a coating liquid (nano-dispersion) with a considerably smaller particle size of ~100 nm. Successful application of such layers was accomplished by means of an innovative coating method, which involves the deposition of a hybrid polymer/GDC membrane by spin-coating and subsequently burning out the polymer part. This method could prevent unwanted infiltration of the nanoparticles into the pores of the anode substrate and the formation of inhomogeneous layers on the relatively rough substrate surface. Results from leakage tests (with air) confirmed that the GDC thin films possess a very low number of defects after sintering at 1400°C and with an average value of 10^{-4} mbar.l.s^{-1}.cm^{-2} the gas-tightness of the first prototypes is promising. In the second part, PVD is studied as an alternative for the wet-coating method and the first results of GDC layers deposited by magnetron sputtering are reported. Significant progress in the formation of dense layers has been obtained by the introduction of additional bias power in the sputtering process, but compatibility issues with our regular anode substrate seem to hinder the formation of layers with a sufficient gas-tightness until now.

INTRODUCTION
    Initially, SOFCs were often prepared with the electrolyte layer as supporting material. As a material, typically yttria-doped zirconia (8YSZ) has been applied. Of course, resistance losses were significant, because of the required large thickness of the electrolyte layer. More recently, by using the anode as the support, the electrolyte thickness has been reduced to ~10 μm, which gives a decrease in ionic resistance by more than one order of magnitude across the electrolyte [1-3].
    Since the ion conductivity through the electrolyte membrane is inversily proportional to the layer thickness, further reducing the thickness of the 8YSZ electrolyte would significantly improve the power density and allow a reduction of the cell operating temperature to the range 600-700°C for 1 μm thick layers and even lower for a layer thickness <1 μm. Alternatively, for cell operation at a temperature below 600°C, materials with a larger conductivity such as doped ceria (e.g. GDC, SDC) and electrolytes based on bismuth oxide and LaGaO₃ perovskite have been frequently mentioned as candidate materials [4,5]. In this paper, we investigate the formation of thin film electrolytes with a thickness of ~1 μm and we selected GDC as an electrolyte material.
    A number of thin film deposition methods have already been proposed in the literature, including particularly pulsed laser deposition (PLD), physical vapour deposition (PVD) and chemical vapour deposition (CVD) (e.g. reference Nr.6). The deposition processes are however frequently carried out on dense substrates such as Si-wafers or sapphire substrates or specially designed porous

substrates rather than practical porous electrodes. The purpose of our research is to study the deposition of thin film GDC electrolyte layers on a regular SOFC substrate, consisting of anode support and anode functional layer (AFL), which is typically characterized by a relatively rough surface, a high porosity and a large pore size.

In the first part of the article, a study of the wet chemical deposition of GDC (20% Gd) thin film electrolyte layers is presented. Previously, impressive results have already been achieved at our institute with such a method, including the preparation of 5x5 cm^2 cells composed of our regular Ni/8YSZ anode substrate, a 1 μm thick 8YSZ electrolyte layer and a LSCF cathode, which show a power output >1 W/cm^2 at 650°C [7]. In this paper, spin-coating, which is an established wet coating technique with a potential for large-scale production, is applied as a coating technique for the GDC layers. In the second part of the article, the deposition of GDC thin films with physical vapour deposition is evaluated. Continuing work on the development and electrochemical characterization of complete cells is ongoing and will be presented in a forthcoming publication.

EXPERIMENTAL

1. Anode Substrate

The substrate preparation method includes a warm pressing method (Coat mix©) to prepare the support plate and subsequently the anode functional layer (AFL) is coated by vacuum slip casting. In both steps, commercial NiO (Mallinckrodt Baker) and 8YSZ powders (Support: Unitec Ceramics, AFL: Tosoh) are used. Presintering of the support plate is done at 1230°C for 3h; the AFL is pre-sintered at 1000°C for 2h, unless stated otherwise.

The finished anode substrate has a dimension of ~25x25 cm² and a thickness of 1 mm or 1.5 mm. The characterization of the substrate surface has been previously described e.g. in reference Nr.7. Basically, in both SEM (Zeiss Ultra) and 3D-profilometry (Cybertechnologies CT 350S), the surface shows a significant roughness, characterized by strongly curved areas. Roughness data in Table 1 of 3 different 7.5x7.5 cm² substrates (Fig. 1a), which were cut from the same 25x25 cm² plate, confirm this. It is assumed that the roughness stems from the press-forms in the manufacturing step of the support.

As shown in Fig.1b, a typical macroporous structure can be recognized for the AFL, with a particle size of 200–400 nm and an average pore size of 200–300 nm. The thickness of the AFL is in the range 5–10 μm.

(a)   (b)

Fig. 1a. Substrate consisting of NiO/8YSZ anode support plate and anode functional layer (size 7.5 x 7.5 cm²). Fig. 1b. Surface micrograph of the AFL (bar = 200 nm).

Table 1. Surface roughness characterization of the anode substrate (support plate + AFL)

Substrate	R_a (µm)	R_q (µm)	R_z (µm)	R_{max} (µm)
Sample 1	0,508	0,665	3,667	4,846
Sample 2	0,613	0,783	4,125	5,239
Sample 3	0,519	0,678	3,837	5,209

2. Preparation of the coating liquids

The preparation route of the coating liquid is schematically summarized in Figure 2. The coating liquid is a nano-dispersion, made by dispersing a commercial GDC nanopowder (Sigma-Aldrich (20% Gd)) in a 0.05 M aqueous HNO_3 solution by ultrasonication and subsequent separation of larger agglomerates by centrifugation. As shown in Figure 3, analysis of the particle size distribution gives an average particle size of ~100 nm and the appearance of the liquid was opaque. The particle size was measured with a dynamic laser beam scattering method (Horiba LB-550).

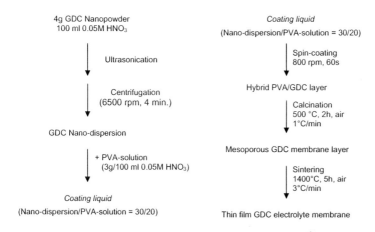

Fig.2. Schematic overview of the preparation route of the porous GDC precursor membrane layer. Final (co-)sintering is carried out at 1400°C after the deposition of the respective layers.

3. Spin coating of thin film GDC layers

To deposit a thin film GDC layer, spin-coating was applied in the first series of coating experiments (Süss MicroTec D80T2 spin-coater). A typical coating step involved dropping 10 ml of the coating liquid onto an anode substrate with a size of 7.5x7.5 cm², which was hold by a vacuum chuck, and spinning the substrate at 800 rpm during 1 minute. To prevent strong infiltration of the GDC particles during the coating step, polyvinyl alcohol was added to the coating liquid (PVA, Merck, 60.000 g/mol). As could be seen from the dried coating liquids, PVA forms a hybrid polymeric membrane layer with the GDC nanoparticles distributed in it, thus preventing infiltration of the nanoparticles during the coating step. After the coating step, the substrate was fired in a conventional furnace at 500°C in air. Then, the coating and calcination step was repeated twice.

Finally, the mesoporous precursor GDC layer was sintered at a temperature of 1400°C in air. A well-known effect is that at this temperature the AFL is also (partially) sintered. The required anode porosity for cell operation is however obtained, when NiO is reduced to metallic Ni during cell operation in reduced atmosphere. In order to characterize the gas-tightness of the thin film GDC layer accurately and to exclude the influence of the sintered AFL, the samples described in this work were reduced at 900°C in Ar/H_2 (4% H_2).

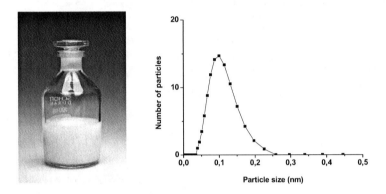

Fig.3a. GDC coating liquid. Fig.3b. Particle size distribution of the GDC coating liquid used in this work.

4. Physical vapour deposition of GDC layers

In a second series of coating experiments, GDC layers were applied by a physical vapour deposition (PVD) method, using a commercial PVD cluster system CS 400 ES (Von Ardenne Anlagentechnik, Germany). In all the experiments, the GDC layer was deposited by reactive magnetron sputtering, using a metallic cerium gadolinium alloy target with a nominal alloy composition of 80 at.% Ce and 20 at.% Gd (99.7 % purity of the alloy) under an O_2/Ar mixture. The base pressure in the process chamber was 10^{-6} Pa (10^{-8} mbar). The specific target power density during deposition ranged from 1–2 W/cm^2, leading to a deposition rate in the range of 10 nm per minute. In a number of experiments, additional bias power was applied to the substrate holder, leading to an additional Ar-ion bombardment of the substrate and the GDC layer.

Pre-sintering of the AFL was done at 1000°C or at 1400°C, in order to enhance the mechanical strength of the AFL. Prior to the deposition process, the substrate surfaces were cleaned by supersonic cleaning. After drying, the substrates were sputter-etched and subsequently coated. The substrates were heated at a rate of 3 K per minute to the set point temperature of 800 °C.

5. Characterization of the GDC thin films

Characterization of the GDC thin films was done by FEG-SEM (Zeiss Ultra 55) and leak testing. SEM characterization included (1) breaking the original sample into a piece of approximately 1 cm^2, (2) mounting it on a conventional sample holder for surface or fracture surface SEM images, (3) sputtering an approximately 2 nm thick Pt layer, (4) applying a conductive copper tape. The gas-tightness of the GDC layers was measured using a commercial leak testing system (dr.wiesner INTEGRA). The dimension of the test area inside the sealing was 4x4 cm^2, which is the same as in our standard leak tests for small 5x5 cm^2 half-cells, and air was the test gas in all tests. It should be noted that the treshold specific leakage for conventional thicker electrolyte layers is set at 2.10^{-5} mbar.l.s^{-1}.cm^{-2} in our institute. This specific leakage is normalised to the measuring area and to a pressure difference of 100 hPa, which is typical for an SOFC stack.

RESULTS

1. Spin coating of thin film GDC layers

Figures 4a and 4b show SEM fracture micrographs of a mesoporous GDC precursor layer, which was synthesized with the 100 nm nano-dispersion and fired at 500°C, as previously described in Figure 2. Micrograph 4b was taken in the back-scatter mode, which gives an improved contrast between layers with a different pore size or porosity. In this micrograph, the anode support plate, the AFL and the mesoporous GDC precursor layer can be clearly recognized and the GDC layer is also clearly visible as a brighter film, due to its much smaller pore size. Previous research on mesoporous membranes for gas separation applications has indicated that such a mesoporous GDC material shows a pore size of ~5 nm after firing at 500°C [8].

In the detail fracture micrographs 4c and 4d, it is confirmed that infiltration of GDC particles into the macropores of the AFL, which have a pore diameter of several hundred nanometer, could be prevented. In Figure 4d, two separation lines are also visible in the GDC layer, which mark an area with a higher concentration of particles at the separation between each single layer. Further, it appears that a thickness of ~1 μm was obtained for a single layer, made by one coating-calcination step.

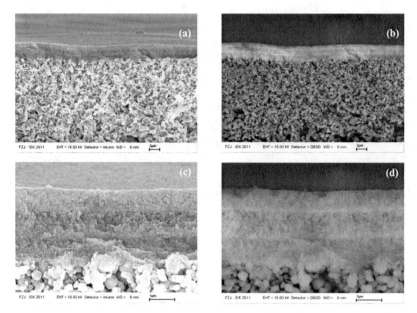

Fig.4. Mesoporous GDC precursor layer, obtained by dip-coating the 100 nm nano-dispersion (3 coating steps, firing at 500°C). Fig.4a,4c. Fracture micrographs. Fig.4b,4d. Fracture micrographs in the back-scattering mode, showing the individual layers. ((a,b) bar = 2 μm, (c,d) bar = 1 μm)

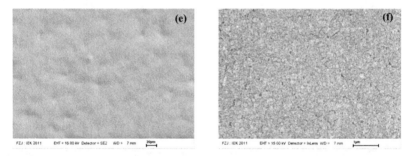

Fig.4e. Overview surface micrograph. Fig. 4f. Detail surface micrograph. ((e) bar = 200 nm, (f) bar = 1 μm)

The overview and detail surface micrographs after firing at 500°C confirm the formation of a homogeneous GDC precursor layer over the wavy substrate surface (4e) with a typical fine mesoporous structure (4b).

Densification of the precursor layer during firing at 1400°C is shown in micrographs 5a-5h. From the fracture micrographs 5a-5d, it seems that a dense homogeneous layer with a thickness of ~1–2 μm was obtained. The surface micrograph 5e shows that such a layer is composed of comparatively large grains, with a size clearly exceeding 1 μm, which suggests that some of the grains in the layer should have a flattened shape, with a width that exceeds the thickness of the grain. Further, it is clear that the thickness of the GDC layer corresponds to the thickness of one single grain.

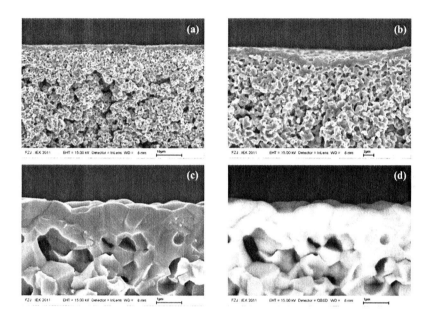

Fig.5. GDC layer after sintering in air at 1400°C and reduction in Ar/H$_2$. Fig.5a-c. Fracture micrographs. Fig.5d. Fracture micrographs in the back-scattering mode. ((a) bar = 10 μm, (b) bar = 2 μm, (c,d) bar = 1 μm)

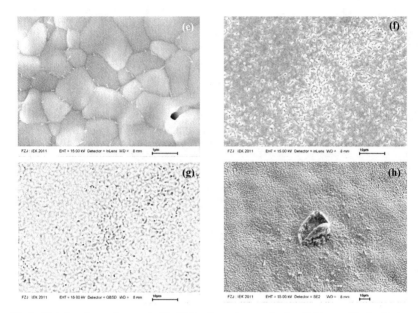

Fig.5e. Detail surface micrograph. Fig.5f. Surface micrograph of untight area. Fig.5g. Backscatter image of this area. Fig.5h. Surface micrograph of large defect. ((e) bar = 1 μm, (f,g,h) bar = 10 μm)

Subsequently, the gas tightness of the sintered GDC layers was investigated in three series of three different samples, which were prepared in the same way. The results of these tests, shown in Table 2, were in agreement with the experimental results discussed in a previous paper, which reported a.o. the deposition of similar thin films consisting of 8YSZ [9]. Basically, the GDC thin films made with the 100 nm nano-dispersion showed comparable leak rate values in the range 5.10^{-5} to 3.10^{-4} mbar.l.s^{-1}.cm^{-2} (average $1.8 \ 10^{-4}$ mbar.l.s^{-1}.cm^{-2}) as 8YSZ thin films previously developed with nano-dispersions with a particle size of 85 nm and 60 nm. The results obtained here confirmed also that an effective coating process has been developed for the reproducible deposition of thin film GDC electrolyte layers with a thickness of 1–2 μm, on a regular anode substrate.

Table 2. Leak test results for 3 series of 3 samples with spin-coated 1–2 μm thick GDC layer

Reproducibility coating series	Specific air leak rate after sintering at 1400°C for 5h (mbar.l.s^{-1}.cm^{-2})		Specific air leak rate after reduction at 900°C in Ar/H$_2$ for 3h (mbar.l.s^{-1}.cm^{-2})
Series Nr.1	Sample 1	8.66 E-05	6.87 E-04
	Sample 2	5.33 E-05	5.36 E-04
	Sample 3	6.07 E-05	6.26 E-04
Series Nr.2	Sample 1	1.44 E-04	7.41 E-04
	Sample 2	1.35 E-04	1.54 E-03
	Sample 3	3.65 E-04	3.38 E-03
Series Nr.3	Sample 1	1.71 E-04	1.47 E-03
	Sample 2	2.38 E-04	1.86 E-03
	Sample 3	3.26 E-04	1.33 E-03

It is also evident from Figures 5f-5h that a further improvement of the gas-tightness is to be expected, when two types of defects can be avoided. First, as shown in Figure 5f and the back-scatter image Figure 5g, the layer contains a number of untight areas. Second, very large defects are present in the layer as shown in Figure 5h. The origin of these defects is now investigated in order to avoid them in our future coating experiments. In analogy with the previous results obtained for 8YSZ thin films, further improvement of the gas-tightness can also be expected by the introduction of additional sol-gel coatings [9].

2. Physical vapour deposition of thin film GDC layers

Figures 6a-6d show detail SEM fracture and surface micrographs of sputtered GDC layers, deposited on a substrate with the AFL pre-sintered at 1000°C (a,b) and at 1400°C (c,d). In these experiments, no bias power was applied and the substrate was porous (a,b) or dense (c,d). Apparently, very untight layers with a characteristic columnar structure were obtained, irrespective of the porosity of the surface.

In a second series of coating experiments, the application of 400W bias power yielded a significantly improved result for samples with the AFL pre-sintered at 1000°C and at 1400°C and based on the SEM pictures a dense layer was obtained (Figures 6e-6h).

Then, these samples were further treated in Ar/4%H$_2$ at 900°C for 3 hours. Figure 7a shows that in case of a pre-sintering temperature of 1000°C for the AFL, serious delamination of the sputtered GDC layer occured. As shown in Figure 7b, an improved result was obtained in case of a pre-sintering at 1400°C, but it is clear that also here an untight layer was obtained, which shows serious cracks.

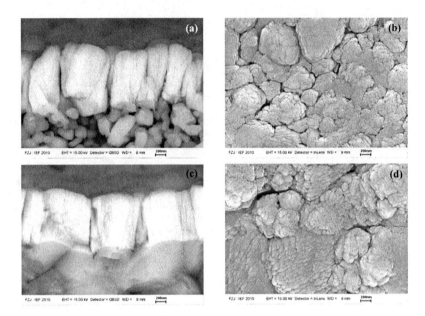

Fig.6. Fracture and surface micrographs in the back-scattering mode of sputtered GDC layers. Fig.6a, 6b. Experiment with AFL pre-sintered at 1000°C and no bias power. Fig.6c,6d. Experiment with AFL pre-sintered at 1400°C and no bias power. ((a-d) bar = 200 nm)

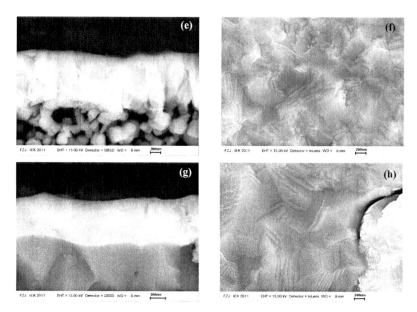

Fig.6e,6f. Experiment with AFL pre-sintered at 1000°C and 400W bias power. Fig.6g,6h. Experiment with AFL pre-sintered at 1400°C and 400W bias power. ((e-h) bar = 200 nm))

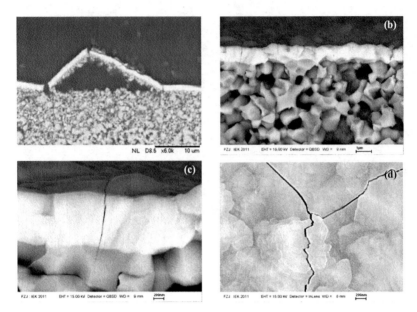

Fig.7. Sputtered GDC layer (400W bias power) after reduction in Ar/H$_2$. Fig.7a. Fracture of sample with AFL pre-sintered at 1000°C. Fig.7b-7d. Fracture and surface of sample with AFL pre-sintered at 1400°C.

Based on the previous observations, it is clear that significant progress has been achieved in the development of thin film GDC electrolyte layers. A practical and scalable wet coating process has been developed, for the deposition of gas-tight thin film GDC layers on a regular anode substrate. Such thin films have a thickness of 1–2 μm and a leak rate for air in the range 5.10^{-5} – 3.10^{-4} mbar.l.s^{-1}.cm^{-2} after sintering at 1400°C, which is also the typical firing temperature in a conventional SOFC preparation route. After treatment in a reducing atmosphere, the leak rate was in the range 5.10^{-4} – 3.10^{-3} mbar.l.s^{-1}.cm^{-2}. This achievement is mainly based on the selection of an additive (polyvinyl-alcohol (PVA)) which gives in combination with the GDC nano-dispersion a perfect coating behaviour in the spin-coating step. The efficiency of the developed coating process was evident from SEM analysis of the calcined mesoporous precursor layer, which indicated that unwanted infiltration of the nanoparticles into the pores of the AFL could be avoided and very homogeneous precursor layers were formed.

Coating experiments with physical vapour deposition (PVD) on the other hand did not result in homogeneous and crack-free layers on our regular substrate yet. From the first experiments, it can be concluded that the application of bias power during the sputtering process significantly enhances the densification of the GDC thin film. After treatment in a reducing atmosphere, however, delamination and cracking of the thin film was observed.

ACKNOWLEDGEMENT
Robert Mücke and Sebastian Vieweger are acknowledged for their assistance with profilometry measurements and leak testing.

REFERENCES
[1] E.D. Wachsman, S.C. Singhal, Solid oxide fuel cell commercialization, research and challenges, American Ceramic Society Bulletin, Vol. 89 (2010) No.3, p. 22-32
[2] O. Yamamoto, Solid oxide fuel cells: fundamentals and prospects, Electrochimica Acta, 45 (2000) 2423-2435
[3] B.C. H. Steele, A. Heinzel, Materials for fuel-cell technologies, Nature, Vol. 414 (2001) 345-352
[4] E.D. Wachsman, K.T. Lee, Lowering the Temperature of Solid Oxide Fuel Cells, Science, Vol. 334 (2011) 935-939
[5] M. Cassir, E. Gourba, Reduction in the operating temperature of solid oxide fuel cells – potential use in transport applications, Ann. Chim. Sci. Mat, 2001, 26 (4) 49-58
[6] D. Beckel, A. Bieberle-Hütter, A. Harvey, A. Infortuna, U.P. Muecke, M. Prestat, J.L.M. Rupp, L.J. Gauckler Thin films for micro solid oxide fuel cells, Journal of Power Sources 173 (2007) 325-345
[7] T. Van Gestel, H. Feng, D. Sebold, R. Mücke, N. Menzler, H.P. Buchkremer, D. Stöver, Nanostructured solid oxide fuel cell design with superior power output at high and intermediate operation temperatures, Microsystem Technologies 17 (2) (2011) 233-242
[8] T. Van Gestel, Microporous membranes for H_2 separation – State of the art and future prospects, lecture at FZJ Membrane Workshop, October 5-6, 2011
[9] T. Van Gestel, D. Sebold, H.P. Buchkremer, D. Stöver, Assembly of 8YSZ nanoparticles into gastight 1–2 μm thick 8YSZ electrolyte layers using wet coating methods, Journal of the European Ceramic Society 32 (2012) 9-26

Author Index

Author Index